Lecture Notes in Electrical Engineering

Volume 945

The book series *Lecture Notes in Electrical Engineering* (LNEE) publishes the latest developments in Electrical Engineering - quickly, informally and in high quality. While original research reported in proceedings and monographs has traditionally formed the core of LNEE, we also encourage authors to submit books devoted to supporting student education and professional training in the various fields and applications areas of electrical engineering. The series cover classical and emerging topics concerning:

- Communication Engineering, Information Theory and Networks
- Electronics Engineering and Microelectronics
- Signal, Image and Speech Processing
- Wireless and Mobile Communication
- Circuits and Systems
- Energy Systems, Power Electronics and Electrical Machines
- Electro-optical Engineering
- Instrumentation Engineering
- Avionics Engineering
- Control Systems
- Internet-of-Things and Cybersecurity
- Biomedical Devices, MEMS and NEMS

For general information about this book series, comments or suggestions, please contact leontina.dicecco@springer.com.

To submit a proposal or request further information, please contact the Publishing Editor in your country:

China

Jasmine Dou, Editor (jasmine.dou@springer.com)

India, Japan, Rest of Asia

Swati Meherishi, Editorial Director (Swati.Meherishi@springer.com)

Southeast Asia, Australia, New Zealand

Ramesh Nath Premnath, Editor (ramesh.premnath@springernature.com)

USA, Canada

Michael Luby, Senior Editor (michael.luby@springer.com)

All other Countries

Leontina Di Cecco, Senior Editor (leontina.dicecco@springer.com)

**** This series is indexed by EI Compendex and Scopus databases. ****

Jaco du Preez · Saurabh Sinha

State-of-the-Art of Millimeter-Wave Silicon Technology

Jaco du Preez
University of Johannesburg
Johannesburg, South Africa

Saurabh Sinha
University of Johannesburg
Johannesburg, South Africa

ISSN 1876-1100 ISSN 1876-1119 (electronic)
Lecture Notes in Electrical Engineering
ISBN 978-3-031-14657-2 ISBN 978-3-031-14655-8 (eBook)
https://doi.org/10.1007/978-3-031-14655-8

This Springer imprint is published by the registered company Springer Nature Switzerland AG
The registered company address is: Gewerbestrasse 11, 6330 Cham, Switzerland

Contents

Chapter 1
Evolution of Millimeter-Wave Silicon Technology

Innovation and evolution are paramount in a world with an insatiable requirement for higher bandwidth, more data, and ubiquitous connectivity. Standards to define mobile backhaul, small cell, fixed broadband and next-generation mobile networking applications are deliberately moving into millimetre-wave (mm-wave) bands. The 802.11 Wireless Local Area Networking (LAN) Working Group specifies physical layer (PHY) operation in the 60 GHz band for 802.11ad WiFi networks and 802.11ay WLANs [1–3]. Additionally, 5G New Radio (NR) applications will utilise the 26 MHz, 28 MHz and 39 MHz bands. 5G NR is a vital enabling technology for Fourth Industrial Revolution (4IR) applications such as massive Internet of Things (IoT), industrial 5G networks and C-V2X (cellular vehicle-to-everything) capabilities for autonomous vehicles [4–6]. The 70–86 GHz E-band region will primarily serve satellite communications and wireless backhaul [7]. Wireless links for wearable devices is another potential application. Here, standards like WirelessHD, IEEE 802.15.3c and ECMA-387 provide the framework for connecting several popular devices that we use each day—smartphones, fitness trackers and wireless headsets [8].

These frequency allocations represent only a minute portion of applications that will quite likely, eventually, utilise mm-wave bands.

Alongside the evolution of wireless standards, the fundamental technologies accompanying and driving rapid improvements in our products have experienced unprecedented growth in recent decades. Perhaps some of the most significant enhancements have been integrated circuit technology, and this text targets the prominence of Si Complementary Metal Oxide Semiconductor (CMOS) as well as SiGe Bipolar CMOS (BiCMOS) technologies. Arguably two of the most popular choices in semiconductor technology, CMOS and BiCMOS, have gained immense traction due to their low fabrication cost and unprecedented integration capability. In addition to this, the radio frequency (RF) and mm-wave performance of both processes have matured them into a formidable competitor for III–IV (such as GaAs and InP)

J. du Preez and S. Sinha, *State-of-the-Art of Millimeter-Wave Silicon Technology*,
Lecture Notes in Electrical Engineering 945,
https://doi.org/10.1007/978-3-031-14655-8_1

technologies in recent years. With that said, Si devices will not be replacing III–IV devices any time soon, as there are applications in which the latter devices are preferred. GaAs, for example, is unparalleled in terms of output power, provide excellent noise figure characteristics and enable oscillators with superb phase noise performance.

In the last few decades, semiconductor technology has steadily grown in maturity, with silicon transistors able to reach increasingly higher unity-gain frequency.

(f_{max}) values. This has proven to be true for technologies based on both CMOS and SiGe BiCMOS. Higher f_{max} values, in turn, lead to transistors suitable for highly complex integrated circuits operating in millimetre-wave bands. The significant performance gains observed in signal processing and other digital circuits based on silicon technologies serve as an excellent motivator for advancing such technologies, particularly CMOS.

Furthermore, performance metrics of digital circuits such as power consumption and computational speed improve alongside technology scaling. Advances in silicon system-on-chip (SoC) solutions have led to integrated circuits (ICs) performing various RF functions in addition to logic and digital processing, using the same CMOS process. Such high levels of integration are advantageous for larger ICs since SoCs reduce cost through reduced interconnect and layout complexity, lower overall power consumption and improve robustness through self-diagnostic functions and on-chip calibration.

1.1 Millimeter-Wave Wireless Communications

The introductory paragraphs of this chapter briefly note the drive toward mm-wave migration. This section will expand on these ideas and further solidify the increase in operating frequency, providing a precursor to the actual semiconductor requirements that face engineers, standardisation organisations and system designers alike. The next decade will likely experience a thousand-fold increase in data traffic [9, 10]. The microwave band between 300 MHz and 3 GHz will certainly not satisfy the demand for capacity of this magnitude, especially since physical layer technology has reached the Shannon limit [11]. The seemingly obvious route to explore would be to increase system bandwidth, where the less congested mm-wave bands (30 to 300 GHz) come into play.

The short wavelengths, narrow beamwidths and intense interaction with the environment (e.g. oxygen and water absorption) pertinent to mm-wave transmission create as many opportunities as challenges. Smaller wavelengths mean that electronic components and transmission lines become physically shorter. Antennas also become much smaller, resulting in densely packed and spatially efficient integrated circuits (ICs) and transceivers. Moreover, wide bandwidths, the potential for wideband spread-spectrum (which generally reduces multipath) and excellent resistance to jamming and interference contribute significantly to the case for mm-wave transmission. One genuine challenge is the precision manufacturing required for mm-wave components, inevitably leading to lower yield and higher cost. Moreover, the fact that receiving antennas can physically handle lower energy levels degrades system sensitivity. The range is also limited to no more than a few kilometres at best, especially in bands where atmospheric attenuation is high [12].

1.1.1 Regulatory Challenges and Spectrum Allocation

Regulatory issues with spectrum assignment, especially in the 6 – 100 GHz region, have hampered the migration towards mainstream mm-wave, despite numerous prominent benefits that address requirements of future mobile networks [11]. The range limitations of mm-wave, typically considered a drawback, can be turned into a positive since this promotes geographic spectrum re-use. The U.S. Federal Communications Commission (FCC) has earmarked several bands up to 100 GHz for various applications [13]:

- The 28 GHz band (27.5 – 29.5 GHz) will service local multipoint distribution services (LMDS).
- The 37 GHz band (37 – 38.6 GHz) holds 1 GHz of unlicensed bandwidth between 37.6 and 38.6 GHz.
- The 39 GHz band (38.6 – 40 GHz) contains 1.4 GHz contiguous spectrum and is best suited for wireless backhaul. Currently allocated licenses are mostly 50 MHz blocks of unpaired spectrum that are licensed as partial economic areas (PEA). This arrangement does not quite fit in with the intended 100 MHz PEA distribution of licenses intended by the FCC. The remaining 39.5 – 40 GHz portion is allocated for various satellite communications applications.
- The 60 GHz V-band (57 – 64 GHz and 64 – 71 GHz) is an ideal candidate for unlicensed short-range applications, and the IEEE 802.11ad specification requires the usage of this band.
- The 70 and 80 GHz E-band (71 – 76 GHz and 81 – 86 GHz) provides 10 GHz of contiguous spectrum and allows backhauling via smaller antennas. These bands are loosely licensed.
- The 90 GHz W-band (92 – 95 GHz) provides 3 GHz contiguous bandwidth, intended for short-range applications, both multipoint and point-to-point.

Spectrum allocation outside of the U.S. is also of interest. China, for example, allocated the 59 – 64 GHz band for short-range wireless links in 2006 [14]. Fixed point-to-point access systems were allocated to the 40.5 – 42.3 GHz and 48.4 – 50.2 GHz bands, while the 42.3 – 47 GHz and 47.2 – 48.4 GHz bands serve mobile point-to-point networks. The IEEE 802.11aj specification details 45 GHz WLAN applications.

In 2018, South Korea auctioned the 28 GHz band to three mobile operators for mm-wave 5G [11]. 5G services were subsequently launched later in 2018, and regulators are considering extending the band to 3 GHz (up from 1 GHz, thereby occupying 26.5 – 29.5 GHz).

The radio regulations body of the International Telecommunications Union (ITU) allocated the 55.78 – 66 GHz, 71 – 76 GHz, 81 – 86 GHz, 92 – 94 GHz, 94.1 – 100 GHz bands globally for fixed wireless access and mobile systems. The ITU also specifies usage of the 64 – 66 GHz band for mobile services outside of aeronautical applications.

Elsewhere, the 59–64 GHz band in Canada is unlicensed and intended for low-power devices. Japan allocated the 59–66 GHz band for unlicensed operation and the 54.25–59 GHz band for licensed usage. Australia allocated the 59.4–62.9 GHz band for unlicensed operation. In Europe, the 59 – 66 GHz band is allocated for mobile service applications, the 59 – 62 GHz for radio local area networks (RLANs), the 58.2 – 59 GHz and 64 – 65 GHz bands for radio astronomy, and the 62 – 63 GHz and 63 – 64 GHz bands for road transportation informatics as well as broadband mobile services.

The complex and technical nature of future wireless networks creates many challenges for regulators. 5G radios utilising intricate beamformed transmissions, for example, add a layer of difficulty to regulation since geographical assignments must be solved in parallel.

1.1.2 Semiconductor Requirements by Application Area

Next-generation transmission networks are highly dependent on the semiconductor technology that drives them. The need for increased throughput, more extended range, availability of service, spectral efficiency, lower power consumption (and therefore, longer battery life) and reduced cost is what drives semiconductor technology development. Additionally, utilisation of frequencies in the millimetre-wave range and, in particular, above 90 GHz is a priority due to the many associated benefits [15]. These performance requirements, such as output power, linearity, noise figure, bandwidth, frequency, packaging, baseband performance, phase noise and integration level, translate to semiconductor devices.

Figure 1.1 shows a generic block diagram of a wireless transceiver, containing some of the significant components covered throughout this book.

The focus of this book is mainly on the components highlighted within Fig. 1.1 The Analogue RF portion consists of mixers, phase-locked loops (PLLs), power amplifiers (PAs), low-noise amplifiers (LNAs) and several passive circuits such as filters, diplexers and power combiners. The Analogue Front End section contains the interface between the digital and analogue parts of the transceiver and primarily involves the analogue-to-digital (A/D) and digital-to-analogue (D/A) converter circuits.

The remainder of the book will not cover the Digital Baseband section in detail, as this is outside of the intended scope. Instead, discussing the integration with the Digital Baseband is of particular interest. Discussions on antennas are also outside of the scope.

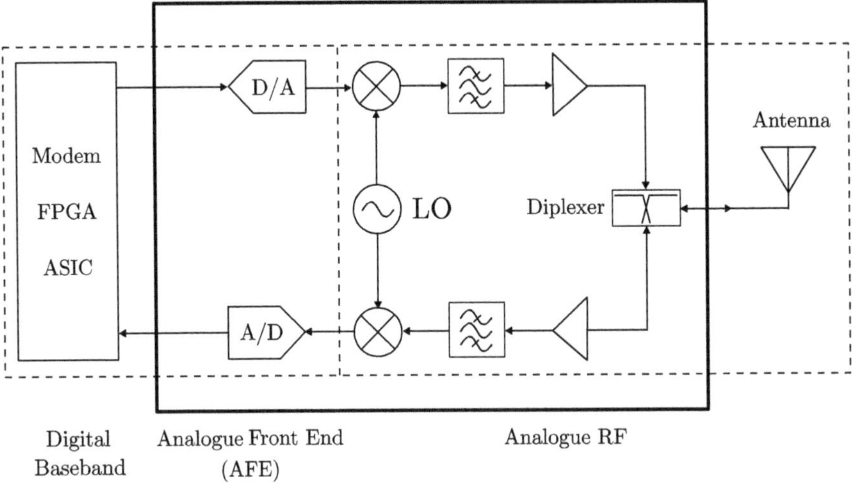

Fig. 1.1 Wireless transceiver block diagram

1.1.2.1 Mobile Backhaul

Allocation of separate uplink and downlink bands in the E-band region (70−90 GHz) enables frequency division duplexing (FDD) for macro-cell mobile backhaul. State of the art E-band circuits are implemented primarily with GaAs devices, and while there is plenty of potential with GaN technologies, it is currently not commercial. Ranges above 2 km and higher-order QAM (constellation sizes of 64 and above) are currently out of reach for SiGe and CMOS devices. These devices operate with lower output power and poorer linearity compared to III–IV alternatives.

V-band backhaul will achieve much lower link distances due to high oxygen absorption at 60 GHz, where the attenuation is roughly 16 dB/km [10]. However, frequency re-use is much easier to implement practically. The single contiguous 57 − 64 GHz band means that FDD and time division duplexing (TDD) is possible. V-band is thus a good option for small cells [16].

Typical required P_{sat} in both E and V-band systems is around 23 dBm, and P_{1dB} of at least 20 dBm is expected, which varies with the employed modulation scheme [17]. Quadrature phase-shift keying (QPSK) requires around 20 dBm, while higher-order schemes such as QAM-256 and above require operation at power backoff in the region of 10 − 12 dBm. Although backoff operation reduces efficiency, this can be acceptable for fixed infrastructure.

Before the low-noise amplifier (LNA), the noise figure contribution by the diplexer, waveguide transitions and other transmission lines ranges between about 7 − 10 dB. Once again, GaAs LNAs are state of the art, but commercial SiGe alternatives achieve around 6−7 dB. One problem at E-band frequencies is that noise figures can vary significantly over frequency, and a delta as high as 4 dB is often observed between −40 °C and +55 °C [18]. A GaAs LNA experiences about a 1.5 dB variation over the same temperature range.

Integrated VCOs are rare, and achieving excellent phase noise performance is mandatory to support higher-order modulation schemes. State of the art SiGe VCOs from Infineon (BGT-70 and BGT-80) typically achieve −82 dBc/Hz phase noise at 100 kHz offset in the 70 – 86 GHz band [19]. In contrast, competitor InGaP oscillators achieve about 10 dB better.

The transceiver output power measured at the antenna port is dependent on the modulation scheme. Generally, output power backoff is less for simpler modulation schemes. The link budget depends on multiple parameters, such as the antenna aperture size, noise figure, system gain and bandwidth and modulation scheme. An E-band QAM-256 transceiver operating will output roughly +10 dBm at the antenna port, and the maximum range is about 1.7 km using a 60 cm antenna [15]. These parameters assume a bandwidth of 500 MHz, system gain of about 60 dB and 99.99% link availability.

Conversely, V-band transceivers can achieve about 1.7 km range using QAM-4 modulation, 50 MHz channel bandwidth (versus 500 MHz) and a 30 cm antenna, provided that an additional PA is used.

1.1.2.2 Small-Cell Mobile Backhaul

Small-cell link distances are in the 100 – 500 m range, which includes street-to-street and rooftop-to-street as well as multi-hop scenarios [20]. Reduced range relaxes the output power requirement significantly, translating to $P_{out} \leq 10$ dBm in E-band, which is easily achievable with CMOS, SiGe BiCMOS or GaAs PAs. The noise figure and phase noise requirements are similar to the macro-cell backhaul scenario. Again, V-band link distances are much shorter, and the high attenuation encourages aggressive frequency re-use, making V-band ideal for small-cell applications. A maximum distance of 500 m translates to 0 – 10 dBm output power for V-band PAs. SiGe BiCMOS and CMOS are likely the preferred technology here since small-cell backhaul involves many short-range links, which means that BOM cost is a priority and integration with digital radios is a bonus.

1.1.2.3 Fixed Broadband

WiGig solutions such as wireless-to-the-home will utilise 60 GHz backhaul, potentially providing an excellent alternative solution for fixed broadband. These systems will operate under the IEEE 802.11ad specification, as well as its follow-up 802.11ay specification [1–3]. In this case, the lower bill of materials (BOM) cost associated with CMOS and SiGe BiCMOS is advantageous and feasible since link distances rarely exceed 3 km. Modest ranges relax PA output power requirements, and the focus can be shifted to linearity performance to support higher-order modulations. The receiver sensitivity requirement is also relaxed.

1.2 Millimeter-Wave Technology Overview

Semiconductor foundries such as the Taiwan Semiconductor Manufacturing Company (TSMC), Samsung Electronics, and ST Microelectronics produce analogue RF components. Fabrication processes typically offered are based on either III–IV (e.g. GaAs and GaN) or silicon (e.g. CMOS and BiCMOS). Processes such as InP and GaN are yet to penetrate commercial markets and thus far only feature prominently in research laboratories.

Traditionally, III–IV technologies (GaAs in particular) have dominated mm-wave applications. They are ideal for front ends, providing excellent phase noise, unmatched output power, and good noisc figure performance.

SiGe and SiGe:C (created by doping Si with small amounts of Carbon and Germanium) have improved drastically in recent years, much of which is credited to the automotive market driving interest in mm-wave. Despite providing lower output power than III–IV devices, SiGe heterojunction bipolar transistors (HBTs) are solid contenders for transceiver front-ends in small-cell applications. It is safe to say that the performance of active SiGe devices is sufficient to be commercially viable in suitable applications, and the focus has moved towards the notorious difficulty of passive devices in Si substrates. Lower Q-factor passives and transmission lines plague Si, creating substantial problems in characterisation at mm-wave bands in addition to complicating tuning.

CMOS provides the lowest cost *if* fabrication volumes are sufficient, and CMOS transistors with much higher f_T compared to just a few years ago, are readily available [10]. Like SiGe processes, the same problems with passive components exist, and CMOS generally underperforms in wireless transceiver systems.

Shrinking transistor dimensions have become faster and generate much less noise, characteristics that favour high bandwidth applications with adequate dynamic range. Primarily, this is due to modest reductions in breakdown and supply voltages in advanced technology nodes. Scaling the emitter and source resistances has proven challenging, and the electron mobility improvements in SiGe processes have just about reached their potential [21]. While most circuit building blocks have benefited immensely from technology scaling, power amplifier (PA) performance has somewhat peaked (especially in CMOS processes). In advanced nodes, reductions in safe operating voltages have been detrimental to achievable PA performance. Voltage-controlled oscillator (VCO) phase noise performance also suffers in deep sub-micron MOSFETs due to the low voltage swing and inherently worse noise performance.

1.2.1 CMOS versus BiCMOS Introductory Comparison

1.2.1.1 Intrinsic versus Extrinsic Transistor RF Performance

Typically, foundries report RF performance at the device level after de-embedding parasitic elements that arise from metal layers [22]. Other methods of de-embedding retain only contacts in their intrinsic transistor models, especially in the case of MOS transistors. Bipolar transistor performance, on the other hand, is affected negligibly by the de-embedding process. Highly scaled CMOS processes with low input capacitance can experience severe effects on f_T from interconnect parasitics [23]. As such, an objective comparison between the two technologies must account for degradation due to the metal wiring stack, which is a strong function of the back-end-of-line (BEOL) of the technology. In reality, the 40 nm and 20 nm CMOS nodes are about where the RF performance peaks. Smaller devices such as FinFETs have much larger parasitic gate capacitance and resistance, heavily degrading RF performance.

1.2.1.2 Intrinsic Device Reliability and Performance

Bipolar transistors inherently offer larger transconductance and lower $1/f$ noise compared to MOS devices. At similar speeds, bipolar devices also come with higher breakdown voltages and better reliability [24]. The main limitation with MOS transistor reliability is hot carrier injection, which degrades g_m and V_T if V_{DS} is sufficiently large. Reducing the supply voltage or lengthening the gate pad will improve lifetime at the cost of RF performance. The consequence is that higher output power (a function of circuit losses, breakdown voltage and maximum tolerable voltage and current swing) can be achieved with BiCMOS.

1.2.1.3 Passives

The fact that BiCMOS technologies benefit from optimised BEOL typically not available in CMOS plays a significant role in the performance of passive devices. In general, BiCMOS passives will perform better at RF and mm-wave frequencies.

1.2.1.4 Fabrication Cost

BiCMOS technologies are typically based on CMOS nodes that are one or two generations old [25]. This implies lower development costs, since older process design kits (PDKs) are generally cheaper. Moreover, manufacturing costs are also reduced seeing that wafers are cheaper. Low and medium production volumes will benefit BiCMOS, but this is a function of the die size. State-of-the-art CMOS nodes achieve much better density, meaning more digital circuitry can be crammed into the same die size, thus negating the cost advantage of BiCMOS.

1.2.2 State-of-the-Art Semiconductor Technologies

A handful of foundries offer analogue mm-wave services. This section briefly compares widely used processes (relevant to mm-wave circuits) for various feature sizes. Note that this is not a comprehensive list since the specifications of some advanced technologies are not publicly available and can therefore not be published here. Nonetheless, this section should give a good indication of the available CMOS and BiCMOS technology nodes.

1.2.2.1 ST Microelectronics

To lower fabrication costs of additional SoI substrate wafers, the 28 nm CMOS FD-SoI process from ST removes several process steps and masking levels from the 28 nm bulk CMOS process. This technology is targeted at low-power mixed-signal designs. The BiCMOS055 process is a high-performance technology, offering state-of-the-art performance in its bipolar and CMOS devices. It features ultra-thick metal layers that enable high-Q inductors. A cheaper alternative is the 130 nm BiCMOS9MW technology, which is based on the 130 nm HCMOS9 technology from ST but with upgraded capabilities for mm-wave systems up to 77 GHz. Table 1.1 summarises process specifications.

1.2.2.2 Ihp

The portfolio offered by Leibniz IHP focuses on RF and digital design, mixed-signal process technology and communication system design. IHP Solutions exists to transfer IHP research activities into commercial products; the company emphasises research and development (R & R&D) activities. State-of-the-art HBTs are provided with cut-off frequencies up to 500 GHz in their high-performance SiGe BiCMOS processes. Their SG25H3 process is a 250 nm technology with higher breakdown voltage HBTs, while their SG13 processes (SG13G2 and SG13S) offer high-performance HBTs. Process specifications are summarised in Table 1.2.

1.2.2.3 Globalfoundries

Globalfoundries is one of the most well-known semiconductor companies with a sizeable portfolio of mm-wave technologies. It offers a range of high-performance technologies, summarised in Table 1.3.

Table 1.1 Summary of ST microelectronics process specifications

Parameter	55 nm BiCMOS055 SiGe	28 nm CMOS FD-SoI	130 nm BiCMOS9MW SiGe
CMOS Gate length	55 nm	28 nm	130 nm
CMOS supply voltage	2.5 V(I/O), 1.0 V (core)	1.8 V(I/O), 1.0 V (core)	2.5 V(I/O), 1.0 V (core)
Metal layers	8	8	6
Thick metals	Ultra-thick top copper (3.0 μm)	Two thick copper layers (0.88 μm)	–
HBT f_T/f_{max}(GHz)	320/370(high-speed) f_T = 180 (medium-voltage)	–	230/280(high-speed) 160/160 (medium-voltage)
CMOS density	700 $k$gates/mm^2(high-speed) 970 $k$gates/mm^2 (high-density)	3000 $k$gates/mm^2	180 $k$gates/mm^2
MOS transistors	Deep n-well, deep-trench isolation Triple-V_t (high/low/standard V_t)		Deep n-well, deep-trench isolation Double-V_t (low-leakage, high-speed)
Temperature range	−40°C to +175°C		
Low-k dielectric	✓	✓	✓
Capacitors	MIM & fringe MIM, MOM	Fringe MOM	MIM
Inductors	Yes, mm-wave	Yes	Yes
Additional features	Embedded memory (single-port RAM/ROM, dual-port RAM)		

1.2.3 Summary of Semiconductor Technology Options

Table 1.4 gives a summary and qualitative comparison of semiconductor technologies [15]. The criteria for comparison include several vital performance specifications:

- Modulation order up to QAM-256, although higher-order QAM may soon be required. IEEE 802.11ax, for example, makes provision for up to QAM-1024, and mm-wave WLAN specifications could soon follow suit [26].
- Minimum channel spacing of 50 MHz as is supported in small-cell 60 GHz applications.
- Bit rates up to 10 Gbps, although future fronthaul applications will likely require more.
- Single-chip transceivers are beneficial and available in SiGe BiCMOS, CMOS and FD-SoI from several manufacturers.
- All technologies mentioned except for InP will likely remain staple in mm-wave systems. InP, due to its cost and complexity, is likely to remain restricted to the defence market.

Table 1.2 Summary of IHP process specifications

Parameter	SG25H3	SG13S	SG13G2
Process node	250 nm	130 nm	
CMOS supply voltage	2.5	1.2, 3.3	
HBT f_T/f_{max}(GHz)	110/180 45/140 25/80	250/340 45/165	300/500 120/330
HBT BV_{CEO}(V)	2.3 5 7	1.7 3.7	1.7 2.5
HBT BV_{CBO}(V)	6 10.5 21	5 15	4.8 8.5
NMOS V_t (V)	0.6	0.71 ($V_{DD} = 3.3$) 0.50 ($V_{DD} = 1.2$)	
PMOS V_t (V)	−0.6	−0.61 ($V_{DD} = 3.3$) −0.47 ($V_{DD} = 1.2$)	
Metal layers (copper)	3 thin (2 μm), 2 thick (3 μm)	5 thin (2 μm), 2 thick (3 μm)	
n+ poly resistors ($\Omega/\blacksquare$)	210	N/A	N/A
p+ poly resistors ($\Omega/\blacksquare$)	280	250	260
High poly resistors ($\Omega/\blacksquare$)	1600	1300	1360
Varactor $C_{\max}/C_{\min}$	3	1.7	
Inductor Q	18, 1 nH @5 GHz 20, 1 nH @10 GHz 37, 1 nH @5 GHz with LBE[a]		
Additional features	PIC,[b] LBE	PIC, LBE, TSV[c]	

[a]Localised backside etching
[b]Photonic integrated circuit
[c]Through-silicon via

Table 1.3 Summary of Globalfoundries process specifications

Parameter	9HP	8XP	8HP	8WL
Process node	90nm	130nm		
CMOS supply voltage	1.2, 1.8, 2.5, 3.3	1.2, 2.5		1.2, 1.8, 2.5, 3.3
TSV		✓	✓	
Metal layers (base/extended)	7/10 (Cu & Al)	5/8 (Cu & Al)		
Thick metals	✓	✓	✓	✓
HBT f_T/f_{max}(GHz)	310/370	250/340	200/265	100/200
HBT BV_{CEO}(V)	2.4 (medium)	3.3	3.5	4.7
MOS devices	Triple well, high/regular-V_t	Triple well, high/regular-V_t	Triple well, high/regular-V_t	Triple well, high/regular-V_t
Thick gate oxide (V)	2.5, 3.3			
Other bipolar devices	–	–	VPNP	–
Resistors	p+ poly, precision p+ poly, high resistance poly, metal, n+ diffusion, silicided poly ballast resistor, low-resistance sub collector	p+ poly, high resistance poly, metal, n+ diffusion, low-resistance sub collector		p+ poly, precision p+ poly, metal, n+ diffusion, silicided poly ballast resistor
Capacitors	VN caps, MIM, dual MIM, high-Q MIM	MIM, dual MIM		
Varactors	NMOS (thick & thin oxide), PMOS (thin oxide), hyper-abrupt	NMOS (thick & thin oxide), hyper-abrupt		
Diodes	PIN, Schottky			
Inductors	Single spiral, series/parallel spiral, symmetrical			
Series/parallel spiral	✓	✓	✓	✓
Symmetrical	✓	✓	✓	✓
Transmission lines	RF wire, coupled wire, coplanar waveguide, mm-wave features (bends, tees, stubs)			

Table 1.4 Millimeter-wave technology comparison summary

Technology	BOM cost	Integration	RF Performance	Typical products
GaAs pHEMT[a]	High for RF modules, requires multi-chip modules	Poor integration with logic circuits	Excellent	PA, LNA, complete RF modules
GaN	Technology not matured	Poor integration with logic circuits	Higher P_{sat} but worse linearity vs. GaAs	Future PA, LNA
InP pHEMT	Greater than GaAs alternatives	Poor integration with logic, high complexity RF circuits	Better f_T, f_{max} compared to GaAs	Primarily military applications
SiGe	Lower than competitors, but depends on production volume	Good RF integration, poorly suited for A/D & baseband integration within SoCs	High f_T, f_{max} due to mm-wave optimisation, good phase noise & low-loss BEOL layer build	Medium to high-power mm-wave transceiver ICs
SiGe BiCMOS	Lower than competitors, but depends on production volume, SoCs result in good yield	Excellent integration with many logic blocks, lower power consumption for A/D and D/A conversion	High f_T, f_{max} due to mm-wave optimisation, good phase noise & low-loss BEOL[b] layer build	Medium to high-power mm-wave transceiver ICs, integration with PLLs and A/D converters possible
RF CMOS	Mm-wave expansion of digital CMOS requires advanced lithography and costly NRE	Allows high-density digital alongside RF, SoC integration possible with digital baseband and memory	Transistor f_T, f_{max} capable of mm-wave operation, requires small process nodes with very low breakdown voltages, translating into lower power and efficiency vs. SiGe BiCMOS	Transceivers and A/D, PLL, and VGA blocks possible, an excellent option for lower performance applications
FD-SoI	Capable of lowest supply voltage and power consumption	Integration with high-density SoCs, relatively low power needed for moving across chip boundaries	Substrate engineering improves RF performance, good isolation and linearity, high f_T, f_{max}	High-speed mixed-signal blocks (e.g. SerDes)

[a]Pseudomorhpic high electron mobility transistors
[b]Back End-Of-Line

References

1. Perahia E, Cordeiro C, Park M, Yang LL (2010) IEEE 802.11ad: defining the next generation. In: 7th IEEE Consumer Communications and Networking Conference (pp 1–5). https://doi.org/10.1109/CCNC.2010.5421713
2. Zhou P et al (2018) IEEE 802.11ay-based mmWave WLANs: design challenges and solutions. IEEE Commun Surv Tutor 20(3):1654–1681. https://doi.org/10.1109/COMST.2018.2816920
3. Ghasempour Y, Da Silva CRCM, Cordeiro C, Knightly EW (2017) IEEE 802.11ay: next-generation 60 GHz communication for 100 Gb/s Wi-Fi. IEEE Commun Mag 55(12):186–192. https://doi.org/10.1109/MCOM.2017.1700393
4. Agiwal M, Roy A, Saxena N (2016) Next generation 5G wireless networks: a comprehensive survey. IEEE Commun Surv Tutor 18(3):1617–1655. https://doi.org/10.1109/COMST.2016.2532458
5. Daneshgar S et al (2020) High-power generation for mm-wave 5G power amplifiers in deep submicrometer planar and FinFET bulk CMOS. IEEE Trans Microw Theory Tech 68(6):2041–2056. https://doi.org/10.1109/TMTT.2020.2990638
6. Lu N, Cheng N, Zhang N, Shen X, Mark JW (2014) Connected vehicles: solutions and challenges. IEEE Internet Things J 1(4):289–299. https://doi.org/10.1109/JIOT.2014.2327587
7. Harati P et al (2017) Is E-band satellite communication viable?: advances in modern solid-state technology open up the next frequency band for SatCom. IEEE Microwave Mag 18(7):64–76. https://doi.org/10.1109/MMM.2017.2738898
8. Wang X et al (2018) Millimeter wave communication: a comprehensive survey. IEEE Commun Surv Tutor 20(3):1616–1653. https://doi.org/10.1109/COMST.2018.2844322
9. Rappaport TS et al (2013) Millimeter wave mobile communications for 5G cellular: it will work! IEEE Access 1:335–349. https://doi.org/10.1109/ACCESS.2013.2260813
10. Rappaport TS, Murdock JN, Gutierrez F (2011) State of the art in 60-GHz integrated circuits and systems for wireless communications. Proc IEEE 99(8):1390–1436. https://doi.org/10.1109/JPROC.2011.2143650
11. Uwaechia AN, Mahyuddin NM (2020) A comprehensive survey on millimeter wave communications for fifth-generation wireless networks: feasibility and challenges. IEEE Access 8:62367–62414. https://doi.org/10.1109/ACCESS.2020.2984204
12. Wang P, Li Y, Song L, Vucetic B (2015) Multi-Gigabit millimeter wave wireless communications for 5G: from fixed access to cellular networks. IEEE Commun Mag 53(1):168–178. https://doi.org/10.1109/MCOM.2015.7010531
13. Federal Communications Commission (2016) Spectrum Frontiers report and order and further notice of proposed rulemaking: FCC16-89, Washington, DC
14. Haiming W, Wei H, Jixin C, Bo S, Xiaoming P (2014) IEEE 802.11aj (45GHz): a new very high throughput millimeter-wave WLAN system. China Commun 11(6):51–62. https://doi.org/10.1109/CC.2014.6879003
15. Rüddenklau U et al (2016) mmWave semiconductor industry technologies : status and evolution. Sophia-Antipolis, France
16. Bojic D et al (2013) Advanced wireless and optical technologies for small-cell mobile backhaul with dynamic software-defined management. IEEE Commun Mag 51(9):86–93. https://doi.org/10.1109/MCOM.2013.6588655
17. Camarchia V, Quaglia R, Piacibello A, Nguyen DP, Wang H, Pham AV (2020) A review of technologies and design techniques of millimeter-wave power amplifiers. IEEE Trans Microw Theory Tech 68(7):2957–2983. https://doi.org/10.1109/TMTT.2020.2989792
18. Varonen M, Kärkkäinen M, Kantanen M, Halonen KAI (2008) Millimeter-wave integrated circuits in 65-nm CMOS. IEEE J Solid-State Circuits 43(9):1991–2002. https://doi.org/10.1109/JSSC.2008.2001902
19. Infineon Technologies AG (2013) One-Chip packaged RF solution for E-band radio, product brief
20. Wells J (2009) Faster than fiber: the future of multi-G/s wireless. IEEE Microwave Mag 10(3):104–112. https://doi.org/10.1109/MMM.2009.932081

21. Pawlak A et al (2015) SiGe HBT modeling for mm-wave circuit design. Proceedings of the IEEE Bipolar/BiCMOS circuits and technology meeting 2015:149–156. https://doi.org/10.1109/BCTM.2015.7340560
22. Chevalier P et al (2018) SiGe BiCMOS current status and future trends in Europe. In: IEEE BiCMOS and Compound Semiconductor Integrated Circuits and Technology Symposium (pp 64–71). https://doi.org/10.1109/BCICTS.2018.8550963
23. Schmid RL, Ulusoy AC, Zeinolabedinzadeh S, Cressler JD (2015) A comparison of the degradation in RF performance due to device interconnects in advanced SiGe HBT and CMOS technologies. IEEE Trans Electron Devices 62(6):1803–1810. https://doi.org/10.1109/TED.2015.2420597
24. Garcia P et al (2008) Will BiCMOS stay competitive for mmW applications? In: IEEE Custom Integrated Circuits Conference (pp 387–394). https://doi.org/10.1109/CICC.2008.4672102
25. Voinigescu SP, Shopov S, Bateman J, Farooq H, Hoffman J, Vasilakopoulos K (2017) Silicon millimeter-wave, terahertz, and high-speed fiber-optic device and benchmark circuit scaling through the 2030 ITRS horizon. Proc IEEE 105(6):1087–1104. https://doi.org/10.1109/JPROC.2017.2672721
26. Bellalta B (2016) IEEE 802.11ax: high-efficiency WLANS. IEEE Wirel Commun 23(1):38–46. https://doi.org/10.1109/MWC.2016.7422404

Chapter 2
Millimeter-Wave Silicon Passive Components

Capacitors, inductors and transmission lines are standard components in any mm-wave circuit and often dominate the component count in a typical MMIC. A commercial-off-the-shelf (COTS) mobile phone has about 20 times as many passive components as active devices [1]. Integrating surface-mount passive components on-chip reduces the form factor of the overall design, and it is standard practice in MMIC design.

As transistor scaling improves high-frequency performance, developing passive components to support high-frequency circuits is becoming increasingly challenging. However, one of the essential enabling aspects of transistor scaling at mm-wave is smaller on-chip passives. The use of smaller devices, in turn, leads to the potential for significant savings in the chip area for mm-wave designs.

2.1 Challenges with Implementing Silicon Passives

As transistors operate closer to f_T, issues such as resistive losses and interconnect parasitic capacitance and inductance become significant in monolithic microwave integrated circuit (MMIC) design. Multiple factors contribute here, which the following sections will cover.

2.1.1 *Ground Plane Requirements*

Passive components implemented on-chip require ground planes to complete the current loop. The ideal ground plane has zero associated impedance, forming a perfect current return path and voltage reference for single-ended circuits. In practice, however, neither of these exist. The return path merely approximates a perfect

J. du Preez and S. Sinha, *State-of-the-Art of Millimeter-Wave Silicon Technology*,
Lecture Notes in Electrical Engineering 945,
https://doi.org/10.1007/978-3-031-14655-8_2

ground, and this only when designed for small, near-zero voltage swing. For example, when an AC source drives a transmission line, the ground conductor sits at 0 V, just as Ampere and Gauss' laws predict. Electromagnetic (EM) field leakage beyond the periphery of the ground plane is thus minimised. In the very likely practical case when the ground shield around the centre conductor is imperfect, however, at least some voltage swing will exist in the ground plane.

Consider the transmission line topologies in Fig. 2.1 [2].

When the outer shield (ground path as shown in Fig. 2.1(a)) is small enough so that its cross-section is identical to that of the signal path, then the voltage carried by the ground conductor will be equal to that of the signal conductor. The remaining two transmission lines in Fig. 2.1—coplanar waveguide (CPW) shown in Fig. 2.1(b) and the microstrip line in Fig. 2.1(c)—lie somewhere in between the extreme cases where $V_{GND} = 0$ or $|V_{GND}| = |V_{signal}|$, which is due to the metal–insulator–metal stackup. A commonly used compensation is to enlarge the ground conductor to be about five times wider than the signal path, which minimises the AC voltage swing.

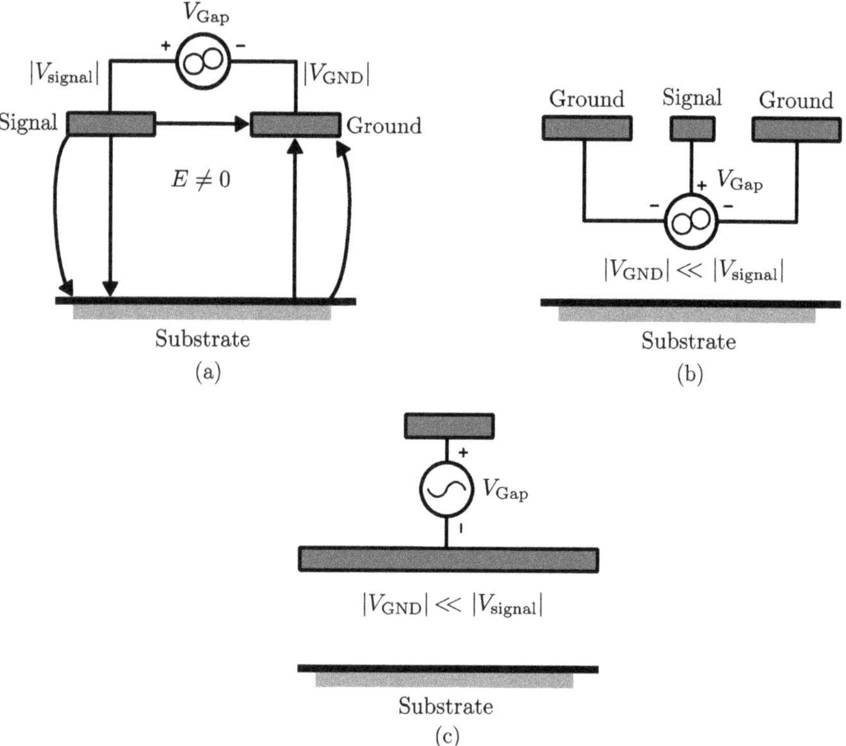

Fig. 2.1 Typical transmission line structures. **a** Balanced, differential **b** Coplanar waveguide **c** Microstrip

2.1.2 Implementing Ground Planes in Silicon MMICs

Implementing true ground references in silicon is challenging since there is no option to connect devices on the top and bottom planes through the substrate (through-wafer connections) [2]. Off-chip ground planes are standard in III-IV technologies and often used as reference and return paths for on-chip components. Ground planes in silicon circuits need to be defined with on-chip metal layers and done explicitly. However, these metal layers are extremely thin and limited to relatively small areas since holes or slots are typically added to circumvent dielectric cracking due to mechanical stress. Another complication is the bondwire connection between on-chip and off-chip ground planes. The combination of parasitic elements due to the thin metal layers and bondwire parasitics required to isolate ground planes is problematic.

2.1.3 Lossy Substrates and Narrow Transmission Lines

Implementing signal and ground planes on a silicon substrate typically results in much thinner signal traces than with other technologies. Semi-insulating substrates typically used in III-IV technologies inherently have low parasitic capacitance, which yields thicker 50Ω signal traces. Additionally, low inductance vias in the substrate connect the chip ground plane to the IC package. Contrast this with silicon MMICs, where a thin dielectric internal layer separates signal and ground, resulting in much larger capacitance. 50Ω signal traces are much narrower, in the order of $6\,\mu$m for a $4\,\mu$m dielectric [2]. Moreover, the Ohmic resistance of the transmission lines is orders of magnitude larger than typical GaAs MMICs.

Another issue with the small gap between signal and ground layers is the Q-factor limitation. Strong capacitive coupling due to the thin dielectric gap between signal and ground planes inhibits the achievable Q-factor.

2.2 Transmission Lines

At mm-wave frequencies, the passive elements required for matching circuits and resonators are small, with inductance values in the low pH$\left(10^{-12}\right)$ range and capacitance values in the fF$\left(10^{-15}\right)$ range [3]. Some on-chip interconnects can be treated as lumped elements at mm-wave frequencies, but the requirement for electrically long connections and matching networks is still prominent [4]. Planar transmission lines are implemented with a metal strip over a ground plane. Intermetal dielectric (IMD) and substrate layers are sandwiched inbetween. The silicon resistivity and operating frequency determine the extent to which the substrate affects the electrical properties of the transmission line. The separation between the top layer conductors and the ground plane should be small, less than $250\,\mu$m at 30 GHz [5].

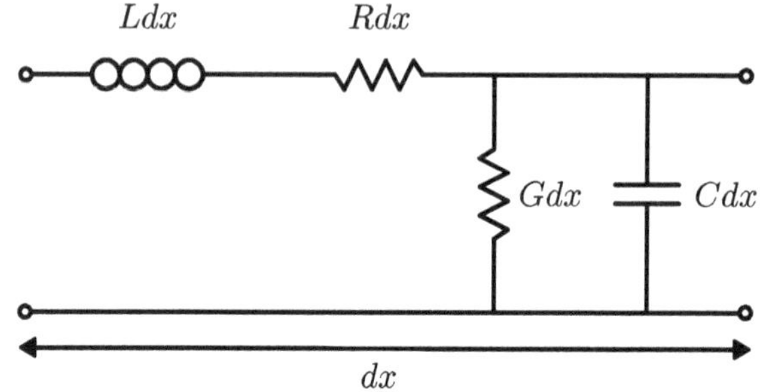

Fig. 2.2 Distributed RLGC transmission line equivalent circuit

Quasi-transverse electromagnetic (quasi-TEM) propagation over transmission lines, particularly microstrip and CPW lines, means that they are inherently length scalable. Additionally, small reactances can be realised precisely. In silicon ICs, however, the skin effect and slow-wave propagation may feature. The frequency-dependent RLGC circuit in Fig. 2.2 characterises a quasi-TEM line.

Real parameters given in (2.1), (2.2), (2.3) and (2.4) also describe quasi-TEM T-lines:

$$Z = \sqrt{\frac{L}{C}} \tag{2.1}$$

$$\lambda = \frac{2\pi}{\omega_0 \sqrt{LC}} \tag{2.2}$$

$$Q_L = \frac{\omega_0 L}{R} \tag{2.3}$$

$$Q_C = \frac{\omega_0 C}{G} \tag{2.4}$$

GaAs T-lines have nearly zero G. In contrast low-resistivity silicon substrates have relatively low capacitive quality factor Q_C), mainly a result of substrate coupling (i.e. G values), which is due to the thin dielectric gap between the signal and ground planes. In T-line structures such as inductors that contain primarily magnetic energy, the inductive quality factor Q_L is the most influential parameter in determining T-line losses.

2.2.1 *Inductive and Capacitive Quality Factors*

A typical application of T-lines in mm-wave circuits is implementing resonant circuits along with parasitic transistor capacitance—in such cases the T-line stores primarily inductive energy (e.g. matching networks). While the total stored energy can be utilised, a more useful parameter to consider is the loss per net reactive energy stored [3]. The net Q-factor is

$$Q_{net} = \frac{2\omega_0}{P_R + P_G}(W_m + W_e) \tag{2.5}$$

where the resonant frequency is ω_0, P_R is the average resistive power dissipated, P_G is the average conductive power dissipated, and W_m and W_e denote average magnetic and electric energy stored, respectively. The individual Q-factors can be rewritten as

$$Q_L = \frac{2\omega_0 W_m}{P_R} \tag{2.6}$$

$$Q_C = \frac{2\omega_0 W_e}{P_G} \tag{2.7}$$

from which two expressions for efficiency can be deduced:

$$\eta_L = 1 - \frac{W_e}{W_m} \tag{2.8}$$

$$\eta_C = \frac{W_m}{W_e} - 1 \tag{2.9}$$

The expression for Q_{net} in (2.5) then simplifies to:

$$\frac{1}{Q_{net}} = \frac{1}{\eta_L Q_L} + \frac{1}{\eta_C Q_C} \tag{2.10}$$

The Q-factor for inductive lines, where $W_m \gg W_e$ and therefore, $\eta_C \gg \eta_L$, will thus reduce to $Q_{net} \approx \eta_L Q_L$, meaning that Q_L contributes almost the entirety of the T-line loss. A short-circuited T-line with $l < \lambda/10$ has $\eta_C > 7\eta_L$, which is an indication of significantly reduced shunt losses. This reduction in shunt losses is vital in silicon substrates where Q_C cannot be neglected.

2.2.2 Microstrip and Coplanar Waveguide Transmission Lines

A typical implementation of silicon microstrip lines utilises the top metal layer for the primary signal route, while the bottom metal layer is the reference ground plane. The metal shield allows virtually no electric energy into the substrate, which means that the oxide loss tangent is the only factor affecting G (the shunt loss), resulting in a relatively high capacitive Q-factor $Q_C \approx 30$ [6].

Implementing microstrip lines in CMOS is challenging because of the narrow separation between the signal and ground layers. This close proximity results in very low L, which reduces Q_L. Coplanar waveguide (CPW) transmission lines are a valuable alternative to microstrip. Ground planes adjacent to the signal line create good coupling to the ground plane and high isolation. The width of the centre trace can be tuned to reduce losses, while the spacing between the signal and ground layers determines Z_0, and can be tuned to achieve the desired Q_L and Q_C. CPW traces can generally achieve much better Q_L values compared to microstrip. Tunability enables low-impedance, high-Q_C or high-impedance, low-Q_L transmission lines, in contrast to microstrip where Q_L and Q_C are consistent regardless of geometry.

CPW losses between 1 to 2 dB/mm at 10 GHz are typical in traditional CMOS processes [7]. Using a costly, high-resistivity silicon-on-insulator (SoI) substrate is one way of reducing losses. In bulk CMOS, however, a practical solution to reduce dielectric losses is by using a metal shield between the metallisation plane and the substrate layer, which inhibits electric field penetration into the substrate. The main drawback of this approach is that eddy currents flow in the shield. Characteristic impedance is inherently low due to the large capacitance per unit length and relatively small permittivity, degrading Q-factors.

A balance between good shielding and practical characteristic impedance ($Z_0 = 50\Omega$) can be achieved by adding a floating shield, resulting in slow-wave CPW (S-CPW) lines [7, 8]. In silicon MMICs, S-CPW lines offer some notable advantages over microstrip and conventional CPW. Microstrip traces, for example, have thinner signal lines and ground planes compared to S-CPW. Traditional CPW traces do not have substrate shields, the lack of which induces higher attenuation. The floating shield is realised with metal strips (not connected to any node in the circuit) underneath the CPW trace. The absence of an electric field tangential to the metal strips provides the shielding property since the shield voltage is zero relative to the CPW. Regardless of the ground and signal conductor dimensions, the shield voltage will remain close to zero. A grounded substrate shield, on the other hand, is affected by parasitic inductances. Figure 2.3 shows the cross-section of an S-CPW trace.

The gap between the signal and ground conductors can be reasonably wide. Keeping the shield strips short in the direction of the current flow ensures that the current induced onto the shield by the signal trace is minimised. One negative characteristic of the S-CPW design is the parasitic capacitance introduced between the signal and ground paths.

2.3 Resistors

Designing MMIC resistors involves either use of the active semiconductor layer or placing resistive metal alloys above the surface [9]. Contact pads are placed on either side of the resistive film in both design methods. These resistors have the same characteristics as surface-mount thin-film resistors. Foundry processes specify their

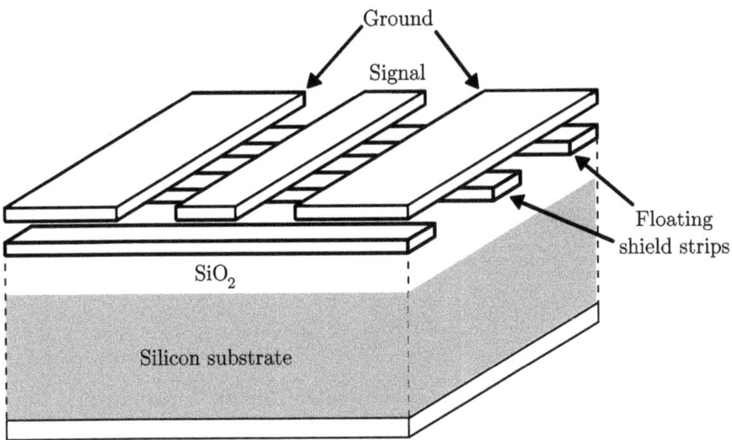

Fig. 2.3 Cross-section of a slow-wave CPW trace

resistive materials in terms of resistance per square (Ω/■). For example, the standard polysilicon material available in the SG13S process from IHP provides 250Ω/■.

Implementing resistors with semiconductor material means that resistance is dependent on the electron mobility in the active layer. As such, resistivity varies with temperature. On the other hand, this option is attractive because much higher Ω/■ can be achieved. Additionally, these resistors are laid out underneath the semiconductor surface, meaning that they do not affect other interconnect routing paths and can be routed without making electrical contact.

2.4 Diodes

Scaling HBTs to obtain very high f_T/f_{max} values degrades the power handling capability since this process reduces the breakdown voltage. SiGe BiCMOS technologies provide the ability to integrate p-i-n (PIN) diodes and Schottky barrier diodes, which enables adding higher power devices to an existing BiCMOS process. Implementing these types of diodes is relatively simple as well since additional masking and processing steps are not required, which helps to keep costs down.

2.4.1 Schottky Barrier Diodes

Schottky diodes are typically encountered in power detector circuits, voltage-controlled oscillators (VCOs), frequency doublers and sub-harmonic mixing blocks. Cutoff frequencies that often reach beyond 1 THz enable BiCMOS Schottky diodes to extend the power handling range of the system significantly, with minimal additional noise and distortion [10]. The off-state capacitance (C_o) and the on-state series resistance (R_s) are two crucial performance-limiting parasitic components present in a Schottky diode. These factors reduce the maximum cutoff frequency (f_c) and hence, reducing their effects is crucial. R_s is extracted experimentally by applying a sufficient bias voltage to the device, while C_s is determined by reverse-biasing the diode and extracting S-parameters [11]. The two parameters then determine the cutoff frequency

$$f_c = \frac{1}{2\pi C_o R_s} \tag{2.11}$$

Additionally, the diode Q-factor is computed from the Y-parameters,

$$Q = \frac{\mathrm{Im}\{-Y_{11}\}}{\mathrm{Re}\{Y_{11}\}} \tag{2.12}$$

Schottky diodes with cutoff frequencies exceeding 1 THz have been reported in CMOS and SiGe processes [12, 13].

2.4.2 *PIN Diodes*

The construction of a PIN diode is quite similar to that of a Schottky diode—the PIN diode uses a resistive contact on the p+ layer as opposed to a Schottky contact. The buried n+ layer naturally arises during fabrication, and the n+ cathode layer is formed through high-dosage implanting into a p-type substrate. The intrinsic layer is then added through epitaxial growth, and a second epitaxial growth process is used to form the p+ anode. The device is isolated from the rest of the substrate.

PIN diodes exhibit similar parasitic resistance and capacitance to Schottky barrier diodes. The main parasitic components are the contact and wiring resistances, down resistance in the intrinsic layer, lateral resistance in the buried layer, and reach-through resistances.

Isolation and insertion losses characterise PIN diode performance. These are analogous to the C_s and R_o values, respectively, of a Schottky diode. Isolation loss indicates how well the diode limits leakage current during the off state. On the other hand, insertion loss is measured in the on state independent of the isolation loss. A representative PIN diode achieves an insertion loss of 0.7 dB and isolation of 11 dB in a 130 nm BiCMOS process [14].

Usage of PIN and Schottky diodes is application dependent. PIN diodes will favour applications that require good isolation with minimal reverse leakage, while Schottky diodes provide much higher switching speeds [12].

2.4.3 Varactor Diodes

Varactor diodes (also called tuning diodes or varicap diodes) are p–n junction devices where the capacitance varies alongside the reverse bias voltage. The layout of the varactor device in bulk CMOS is similar to that of a MOSFET [15]. Connecting the drain and source terminals provides a compact layout. These terminals are then grounded, and the tuning voltage is applied to the gate. MOS varactors are helpful in applications that require tunable capacitors such as voltage-controlled oscillators (VCOs) [16, 17].

2.5 Capacitors

Mm-wave circuits make use of small capacitances in the range of fF range. Interdigital or metal–oxide–metal (MOM) and metal–insulator–metal (MIM) sandwiches are popular methods of implementing MMIC capacitors. Interdigital capacitors are reliant on the fringe capacitance between metal strips, as shown in Fig. 2.4, where l denotes finger length and d is the finger spacing [18].

The separation between fingers (d) is constrained by the minimum gap allotted by the foundry, and they are typically routed on the same layer as transmission lines. MOM capacitors can therefore take advantage of the narrow spacing and thinner dielectric layers associated with scaled technologies [18]. The fringe capacitance is reasonably low, and interdigital structures can only reach about 1 pF, making them useful at mm-wave frequencies. Increasing capacitance density can be achieved by replicating the interdigital finger layout shown in Fig. 2.4 on multiple adjacent metallisation layers. This is referred to as a vertical parallel plate (VPP) or vertical natural capacitor (VNCAP).

Figure 2.5 shows the layout of a MIM capacitor.

MIM capacitors can achieve much higher capacitance values, which results from the large conductive plates forming the capacitive structure and the dielectric separation, further increasing the capacitance [19]. The loss, sizing of the metal plates,

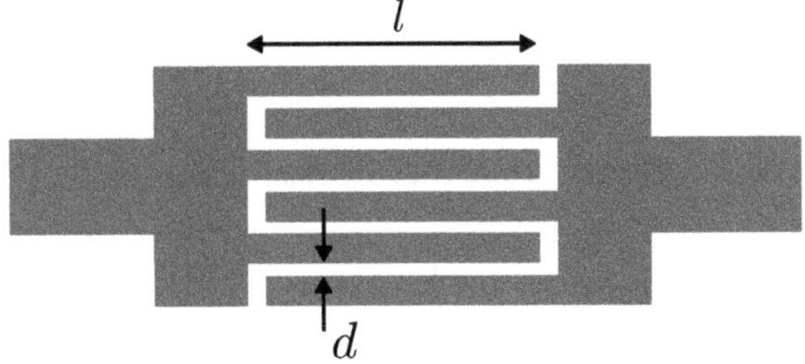

Fig. 2.4 Interdigital capacitor layout

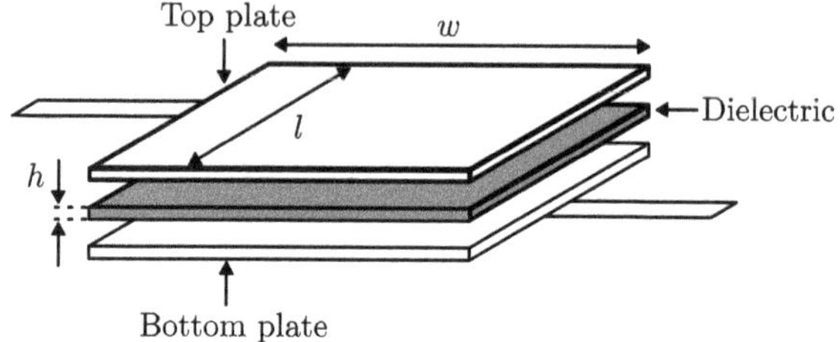

Fig. 2.5 MIM capacitor layout

capacitance and breakdown voltage are functions of the separating dielectric material. Scaled processes such as the SG13G2 from IHP offer MIM capacitors at around $1.0 - 1.5$ fF$/\mu m^2$.

2.6 Inductors

2.6.1 Mm-wave Design Considerations

Silicon-based inductors designed to operate in mm-wave bands are primarily affected by substrate losses, which lowers the attainable Q-factor [20]. Minimising the inductor footprint can help to reduce substrate losses. Reducing the outer diameter and width of the metal traces can accomplish this, which contrasts with RF inductor design methodology where traces are made as wide as possible to reduce series resistance. Operating inductors below their peak-Q frequency is critical to ensuring performance, and this also emphasises the requirement for maintaining a high self-resonant frequency (SRF). Wider spacing between turns is often utilised to improve SRF since it reduces capacitive coupling.

Fortunately, mm-wave inductors are usually in the 100 pH range, which translates to small footprints since they often require only one or two turns. While most digital CMOS processes lack thick top metal layers, a 1 μm metal layer is large enough relative to the skin depth of about 0.3 μm at 60 GHz. Many BiCMOS technologies offer ultra-thick top metal layers of 3 μm and more, enabling simple fabrication of mm-wave inductors.

2.6.2 Planar and 3-D Spiral Inductors

3-D passive components have become quite popular in mm-wave wireless transceiver ICs [21]. The advent of low-temperature co-fired ceramic (LTCC) processes in addition to the system-on-package (SoP) approach to wireless modules has been influential in creating low-cost, high-performance components that offer excellent integration potential [21, 22]. 3-D passive components form an integral part of multilayer,

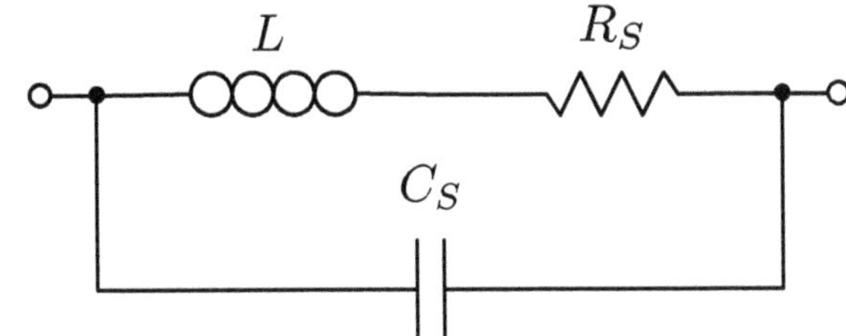

Fig. 2.6 Simplified high-frequency inductor model

high-density designs. Spiral inductors, in particular, provide good Q-factors for a specific inductance value and die area [23].

The inductance in a typical single-layer spiral inductor is tuned by altering the number of lateral turns in the structure. Larger inductance is obtained by adding turns. More turns results in an increase in equivalent series resistance R_S and shunt capacitance C_S. Furthermore, series inductance L is observed as the area of the component increases (proportional to the number of turns)—see the high-frequency inductor model shown in Fig. 2.6 for reference.

This model can be expanded to more accurately represent a spiral inductor implemented in silicon, as Fig. 2.7 shows [24].

The model in Fig. 2.7 uses a grounded substrate, where:

- L_s is the equivalent inductance
- R_s is the series resistance of the metal strip, and its RF performance is determined by the flow of eddy currents
- C_s is a series feedforward capacitance that models the overlap capacitances between the spiral strips and the underpass
- C_{ox} is the oxide capacitance that exists between the substrate and the spiral
- R_{Si} and C_{Si} model the parasitic substrate resistance and capacitance, respectively

The Q-factor of the inductor in Fig. 2.7 is [25]

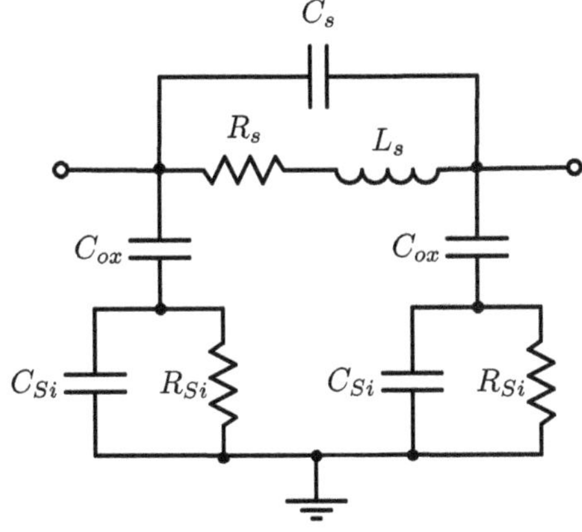

Fig. 2.7 An improved π-model for a spiral inductor implemented in silicon

$$Q = \frac{2\pi f_0 L_S}{R_S} \cdot \frac{R_P}{R_P + \left\{\left(\frac{2\pi f_0 L_S}{R_S}\right)^2 + 1\right\} R_S} \cdot \left\{1 - (C_P + C_S)\left[(2\pi f_0)^2 L_S + \frac{R_S^2}{L_S}\right]\right\} \tag{2.13}$$

where

$$R_P = \frac{1}{(2\pi f_0)^2 C_{ox}^2 R_{Si}} + \frac{R_{Si}(C_{ox} + C_{Si})}{C_{ox}^2} \tag{2.14}$$

and

$$C_P = C_{ox} \cdot \frac{1 + (2\pi f_0)^2 (C_{ox} + C_{Si}) C_{Si} R_{Si}^2}{1 + (2\pi f_0)^2 (C_{ox} + C_{Si}) R_{Si}^2} \tag{2.15}$$

MMIC inductors are fabricated with narrow transmission lines, either used on their own or spirally wound around a centrepiece. The ratio between trace thickness and substrate height determines the inductance per unit length—narrower relative trace thickness will have higher inductance. In addition to the inductance contributed by the trace length, spiral inductors will have additional inductance resulting from additive magnetic fields created by adjacent turns. In essence, mutual inductance between all turns and a large electric field in the middle of the spiral gives good area efficiency, a bonus in a space-constrained MMIC. With that said, turns in the spiral are also coupled capacitively, which means that the spiral structure may not look inductive at all if the operating frequency is sufficiently high.

Spirals are implemented in multiple shapes: square, circular, octagonal and stacked. Figure 2.8 shows a simple layout of a square spiral inductor.

The square corners can be problematic since they lead to stronger capacitive coupling between turns and carry more resistive losses. Circular and octagonal structures provide improved inductive characteristics, but such shapes are typically much more difficult to validate with foundry design-rule checking (DRC) software compared to simple square structures.

Solid ground shields disturb the inductor's magnetic field, inducing an image current into the ground shield. A popular solution is to use patterned ground shields, which increase the resistance to the image current. The slots are effectively open circuits and sever the path of the induced loop current [24]. The shield may be constructed from low-resistivity buried layers, silicided polysilicon or metal [2].

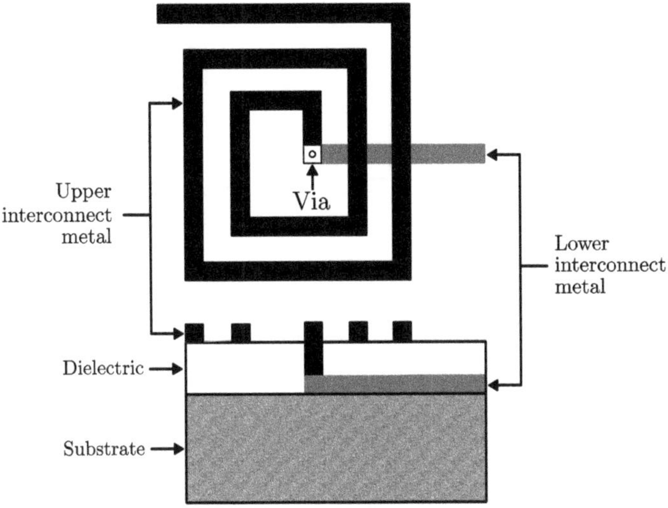

Fig. 2.8 Planar multilayer spiral inductor

2.6.3 *Active Inductors*

The performance of spiral inductors implemented in bulk 65 nm CMOS is heavily affected by loss mechanisms in the substrate. This makes it challenging to realise on-chip high-Q filters. Pepe and Zito report a viable alternative in the form of an active inductor [26]. Figure 2.9 shows a schematic of such an inductor, and its small-signal equivalent is shown in Fig. 2.10.

Fig. 2.9 Active inductor schematic

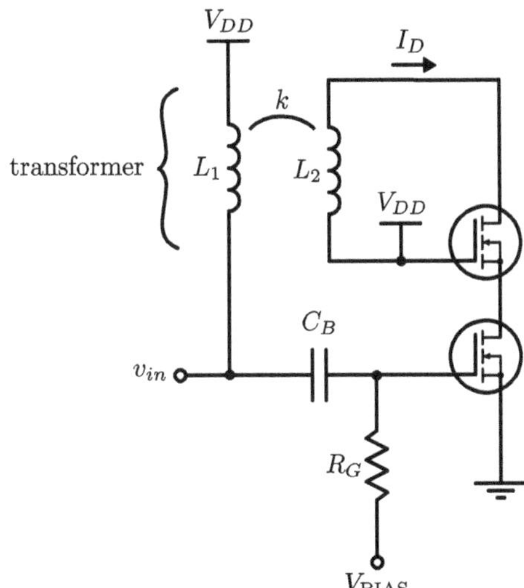

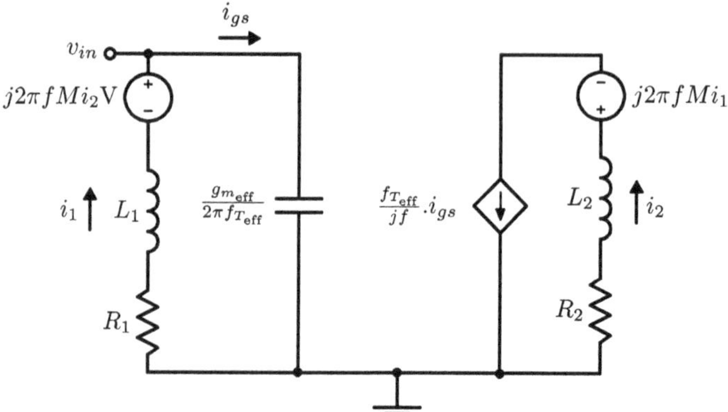

Fig. 2.10 Active inductor small-signal equivalent circuit

The basic idea is to provide the same input impedance as a high-Q inductor by utilising current amplification between the two spirals (L_1 and L_2 in Figs. 2.9 and 2.10) of an integrated transformer [26]. The equivalent circuit in Fig. 2.6 shows that this is achieved by cascading transconductance stages and an on-chip transformer. The input impedance of the circuit in Fig. 2.10 is

$$Z_{in} = \frac{\frac{f_{T_{\text{eff}}}}{jfg_{m_{\text{eff}}}}(R_1 + j2\pi f L_1)}{\frac{f_{T_{\text{eff}}}}{jfg_{m_{\text{eff}}}} - 2\pi f_{T_{\text{eff}}} M + R_1 + j2\pi f L_1} \tag{2.16}$$

where:

- $f_{T_{\text{eff}}}$ is the cascode amplifier cutoff frequency
- $g_{m_{\text{eff}}}$ is cascode amplifier the effective transconductance
- R_1 is the parasitic series resistance of the primary transformer spiral
- $M = k\sqrt{L_1 L_2}$ is the mutual inductance between L_1 and L_2, with k the coupling factor
- f is the operating frequency

From this, the equivalent inductance is

$$L = \frac{\text{Im}\{Z_{in}\}}{j2\pi f} \tag{2.17}$$

and the Q-factor is

$$Q = \frac{\text{Im}\{Z_{in}\}}{\text{Re}\{Z_{in}\}} \tag{2.18}$$

Obtaining the desired Q-factor and inductance is a matter of tuning $g_{m_{\text{eff}}}$ and M. Increasing $g_{m_{\text{eff}}}$ also increases Q. The fabricated inductor reported by Pepe and Zito achieves $Q > 400$ at 50GHz, with an equivalent inductance of 133pH [26]. The downside of this design is the added chip area, but the precision impedance tuning and high-Q inductance can be imperative to the success of many designs.

Only a handful of reported active inductor designs have been implemented in BiCMOS processes. Wei and Qiao utilised an active inductor as part of a 40GS/s A/D converter in a 180nm process [27] and Bosse et al. in a harmonic rejection system fabricated in a 250nm node [28].

2.6.4 Inductor Design Kits

The lossy silicon substrates used in SiGe and CMOS processes limit the Q-factor of on-chip inductors, and mm-wave operation exacerbates this problem. Inductor design requires careful balancing of series, low-frequency conductor losses and shunt, high-frequency substrate losses. The optimum geometry is frequency-dependent, and it is often not ideal to use a standard library of inductors. As such, inductor design kits have become quite popular in recent years.

One such example is the RFIC Inductor Toolkit offered by Dr. Mühlhaus Consulting & Software GmbH [29]. Foundry-validated kits are available for IHP SG25H3/H4, SGB25V, SG13G2/S and Infineon B7HF200, B11HFC PDKs. The toolkit provides an automated design flow and accepts geometry limitations (from the foundry specifications), operating frequency and required inductance. The result is a synthesised inductor, complete with EM simulations and a usable layout. Moreover, the kit is compatible with Keysight Advanced Design System (ADS) Momentum and Sonnet design tools.

2.7 Through-Silicon Vias

In the last two or three decades, technologies to improve semiconductor integration and packaging have primarily focused on 2-D applications. This has been the case for multiple industries: military, bioelectronics and optoelectronics [30]. Applications that incorporate semiconductor technology vary from high-tech military systems down to low-cost and mass-produced consumer products. One of the reasons for this widespread adaptation is the excellent scalability of MOS devices. However, recently developed devices have been shown to deviate from the idealscaling theory established previously [31, 32]. One of the main reasons for this is the challenging nature of scaling supply voltages. Lowering the threshold voltage of a MOS device can generally not be done without increasing subthreshold leakage, seeing that the kT/q value does not scale. Without the ability to scale the threshold voltage, most designs require careful balancing of power and performance.

Two approaches are typically employed towards extracting maximum performance from large-scale ICs while inhibiting power consumption. First, the system architecture can be modified to focus on power consumption. Such modification can be an extremely time-consuming exercise and would perhaps require constant iteration as the design of the IC evolves. Alternatively, restructuring the large-scale integration (LSI) strategy could be a viable option. 3D LSI can lead to improved performance without hampering power consumption [33].

In practice, several methods to implement chip interconnections are used: edge connects, coupling (inductive or capacitive), bond wires, and through-silicon via (TSV). TSVs are especially popular in 3D LSI designs.

Wietstruck et al. report TSVs integrated with the SG13G2/SG13S 130nm technologies from IHP [32]. Integrating TSVs between the Front-end-of-line (FEOL) and BEOL steps allows high aspect ratio TSVs that connect through the wafer, offering sizeable improvements in RF performance.

2.8 Conclusion

This chapter summarised passive components typically used in silicon ICs, along with the performance that can roughly be expected with different foundry processes. While high losses in silicon substrates continue to plague passive component design and naturally inhibit Q-factors, clever innovations like S-CPW transmission lines help combat this. Continued advancement in silicon passives is required for mm-wave CMOS and BiCMOS technologies to increase their market share, given the marked improvements in RF performance in recent years.

References

1. Dunn JS et al (2003) Foundation of rf CMOS and SiGe BiCMOS technologies. IBM J Res Dev 47(2/3):101–138. https://doi.org/10.1147/rd.472.0101
2. Cheung TS, Long JR (2006) Shielded passive devices for silicon-based monolithic microwave and millimeter-wave integrated circuits. IEEE J Solid-State Circuits 41(5):1183–1200
3. Doan CH, Emami S, Niknejad AM, Brodersen RW (2005) Millimeter-wave CMOS design. IEEE J Solid-State Circuits 40(1):144–154. https://doi.org/10.1109/JSSC.2004.837251
4. Long JR, Zhao Y, Wu W, Spirito M, Vera L, Gordon E (2012) Passive circuit technologies for mm-wave wireless systems on silicon. IEEE Trans Circuits Syst I Regul Pap 59(8):1680–1693. https://doi.org/10.1109/TCSI.2012.2206499
5. Bhattacharyya AK (1990) Characteristics of space and surface waves in a multilayered structure. IEEE Trans Antennas Propag 38(8):1231–1238. https://doi.org/10.1109/8.56959
6. Gianesello F et al (2005) State of the art integrated millimeter wave passive components and circuits in advanced thin SOI CMOS technology on high resistivity substrate. Proceedings—IEEE International SOI Conference 2005:52–53. https://doi.org/10.1109/SOI.2005.1563531
7. Franc AL, Kaddour D, Pistono E, Corrao N, Fournier JM, Ferrari P (2009) Miniaturized high performance shielded CPW transmission lines from RF to mm-waves. In: Proceedings of the 39th European Solid-State Device Research Conference (pp 129–133). https://doi.org/10.1109/ESSDERC.2009.5331538

8. Franc AL, Pistono E, Gloria D, Ferrari P (2012) High-performance shielded coplanar waveguides for the design of CMOS 60-GHz bandpass filters. IEEE Trans Electron Devices 59(5):1219–1226. https://doi.org/10.1109/TED.2012.2186301
9. Marsh S (2006) Practical MMIC design. Artech House Inc., Boston, Massachusetts
10. du Preez J, Sinha S (2017) Millimeter-wave power amplifiers. Springer International Publishing
11. Jelenski A, Kollberg E, Schneider MV, Zirath H (1985) Accurate determination of the series resistance of mm-Wave Schottky diodes. In: 15th European Microwave Conference (pp 279–284)
12. Pfeiffer UR, Mishra C, Rassel RM, Pinkett S, Reynolds SK (2008) Schottky barrier diode circuits in silicon for future millimeter-wave and terahertz applications. IEEE Trans Microw Theory Tech 56(2):364–371. https://doi.org/10.1109/TMTT.2007.914656
13. Sankaran S, Kenneth KO (2005) Schottky barrier diodes for millimeter wave detection in a foundry CMOS process. IEEE Electron Device Lett 26(7):492–494. https://doi.org/10.1109/LED.2005.851127
14. Orner BA et al (2006) p-i-n Diodes for monolithic millimeter wave BiCMOS applications. In: International SiGe technology and device meeting (pp 1–2)
15. Shamsadini S, Filanovsky IM, Mousavi P, Moez K (2018) A 60-GHz transmission line phase shifter using varactors and tunable inductors in 65-nm CMOS technology.IEEE Trans Very Large Scale Integr (VLSI) Syst 26(10):2073–2084. https://doi.org/10.1109/TVLSI.2018.2839709
16. Tesson O et al (2013) High-quality varactors and Schottky-diodes in SiGe:C technology for mm-Wave and THz applications. In: Proceedings of the IEEE bipolar/BiCMOS circuits and technology meeting (pp 81–84). https://doi.org/10.1109/BCTM.2013.6798149
17. Nicolson ST et al (2006) Design and scaling of SiGe BiCMOS VCOs above 100GHz. IEEE J Solid-State Circuits 42(9):1821–1833. https://doi.org/10.1109/BIPOL.2006.311135
18. Shi J, Kang K, Xiong YZ, Brinkhoff J, Lin F, Yuan XJ (2010) Millimeter-wave passives in 45-nm digital CMOS. IEEE Electron Device Lett 31(10):1080–1082. https://doi.org/10.1109/LED.2010.2058993
19. Harame DL, Member S, Ahlgren DC (2001) Current status and future trends of SiGe BiCMOS technology. IEEE Trans Electron Devices 48(11):2575–2594
20. Dickson TO, LaCroix MA, Boret S, Gloria D, Beerkens R, Voinigescu SP (2005) 30-100-GHz inductors and transformers for Millimeter-Wave (Bi)CMOS integrated circuits. IEEE Trans Microw Theory Tech 53(1):123–132. https://doi.org/10.1109/TMTT.2004.839329
21. Lee JH et al (2005) Highly integrated millimeter-wave passive components using 3-D LTCC System-on-Package (SOP) technology. IEEE Trans Microw Theory Tech 53(6 II):2220–2229. https://doi.org/10.1109/TMTT.2005.848777
22. Kondratyev V, Lahti M, Jaakola T (2003) On the design of LTCC filter for Millimeter-Waves. IEEE MTT-S International Microwave Symposium Digest 3:1771–1774. https://doi.org/10.1109/MWSYM.2003.1210483
23. Yao T et al (2007) Algorithmic design of CMOS LNAs and PAs For 60-GHz radio. IEEE J Solid-State Circuits 42(5):1044–1056. https://doi.org/10.1109/JSSC.2007.894325
24. Yue CP, Wong SS (1998) On-chip spiral inductors with patterned ground shields for Si-based RF IC's. IEEE J Solid-State Circuits 33(5):743–752
25. Božanic M, Sinha S, Du Plessis M, Müller A (2009) Design flow for a SiGe BiCMOS based power amplifier. Proceedings of the International Semiconductor Conference, CAS 1:311–314. https://doi.org/10.1109/SMICND.2009.5336537
26. Pepe D, Zito D (2014) 50 GHz mm-wave CMOS active inductor. IEEE Microwave Wirel Compon Lett 24(4):254–256. https://doi.org/10.1109/LMWC.2013.2295224
27. Wei HW, Qiao M (2014) A 40GS/s low-power BiCMOS comparator for ultra-high speed ADC. In: 9th international symposium on communication systems, networks & digital sign (pp 934–937)
28. Sy CH, Bosse S, Barth S, Harison SR (2015) Design of an harmonic rejection system using integrated active inductors. In: 21st IEEE International Conference on Electronics, Circuits and Systems (pp 151–154). https://doi.org/10.1109/ICECS.2014.7049944

29. Dr. Mühlhaus and GmbH Consulting & Software (2021) RFIC inductor toolkit: on-chip inductor synthesis. Available: https://muehlhaus.com/products/rfic-inductor-toolkit-for-ads. [Accessed: 08-Oct-2021]
30. Hu S et al (2011) TSV technology for millimeter-wave and terahertz design and applications. IEEE Trans Compon, Packag Manuf Technol 1(2):260–267. https://doi.org/10.1109/TCPMT.2010.2099731
31. Yook JM, Kim YG, Kim W, Kim S, Kim JC (2020) Ultrawideband signal transition using quasi-coaxial Through-Silicon-Via (TSV) for mm-Wave IC packaging. IEEE Microwave Wirel Compon Lett 30(2):167–169. https://doi.org/10.1109/LMWC.2019.2960884
32. Wietstruck M, Marschmeyer S, Wipf C, Stocchi M, Kaynak M (2021) BiCMOS Through-Silicon Via (TSV) signal transition at 240/300 GHz for MM-Wave sub-THz packaging and heterogeneous integration. In: 50th European Microwave Conference (pp 244–247). https://doi.org/10.23919/EuMC48046.2021.9338247
33. Motoyoshi M (2009) Through-Silicon Via (TSV). Proc IEEE 97(1):43–48. https://doi.org/10.1109/JPROC.2008.2007462

Chapter 3
Active Millimeter-Wave Silicon Devices

Semiconductor technology has steadily advanced over the last few decades, and transistors with f_{max} values exceeding $100 - 200$ GHz are now commonplace in multiple processes. As such, a more extensive selection of foundry processes is now available for mm-wave designs. The exponential increase in the capabilities of digital circuits warrants continuous improvements from semiconductor companies to push technology boundaries to stay competitive. Moreover, with the advent of system-on-chip (SoC) solutions, the ability to integrate RF and digital circuitry on the same die has become a reality and a desirable solution in many applications. This is where SiGe BiCMOS and CMOS present a sizeable advantage over III-IV technologies. First, with access to bipolar and CMOS transistors in BiCMOS processes, designers can leverage good RF performance alongside complex digital circuits. Second, the RF performance of bulk CMOS and silicon-on-insulator (SOI) CMOS technologies has improved dramatically, providing another viable alternative to III-IV devices.

Silicon offers several advantageous properties to semiconductor fabrication, leading to its dominance on the IC market. Growing high-quality dielectrics on silicon is relatively painless. The fact that silicon wafers can be grown in large single-crystal structures with close to no defects means that many low-cost ICs can be fabricated on a single wafer. Silicon possesses excellent thermal properties, improving heat handling of dense circuits operating in extreme environments. Moreover, it can handle high mechanical stresses, which eases fabrication, transportation and handling. Controlled doping of n and p-type impurities is possible with high dynamic range, and low-resistance ohmic contacts help reduce parasitics. These properties, alongside a worldwide abundance of silicon, make it extremely attractive from a manufacturing perspective.

One of the main drawbacks of silicon is the low carrier mobility, which determines device speeds. In contrast, the direct-gap operation of GaAs and InP devices facilitates operating at much higher frequencies, stretching into the optical range. Additional performance can also be tapped from III-IV technologies via bandgap engineering, where compositions are tuned to suit a particular application.

J. du Preez and S. Sinha, *State-of-the-Art of Millimeter-Wave Silicon Technology*,
Lecture Notes in Electrical Engineering 945,
https://doi.org/10.1007/978-3-031-14655-8_3

3.1 Bipolar Transistors

This sectiont of the chapter deals with bipolar transistors—in the main bipolar junction transistors (BJTs) and the more modern heterojunction bipolar transistors (HBTs).

A BJT is created by placing two PN junctions back-to-back. BJTs are three-terminal devices used in high-power circuits, amplifiers, switches and high-speed logic. PNP or NPN BJTs are fabricated by changing the sequence of silicon doping, although NPN devices are most often, especially in integrated designs.

3.1.1 Small-Signal Equivalent Circuit

Where both the forward and reverse-biased depletion regions that form between PN junctions have capacitances associated with them. The base-emitter capacitance is denoted by C_π and the base–collector capacitance by C_μ. Integrated bipolar transistors are fabricated on top of a grounded substrate [1]. Figure 3.1a shows the structure of such a bipolar device, followed by its small-signal equivalent circuit in Fig. 3.1b.

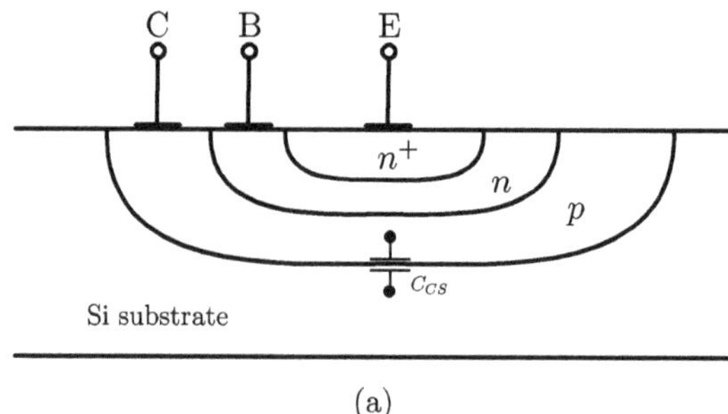

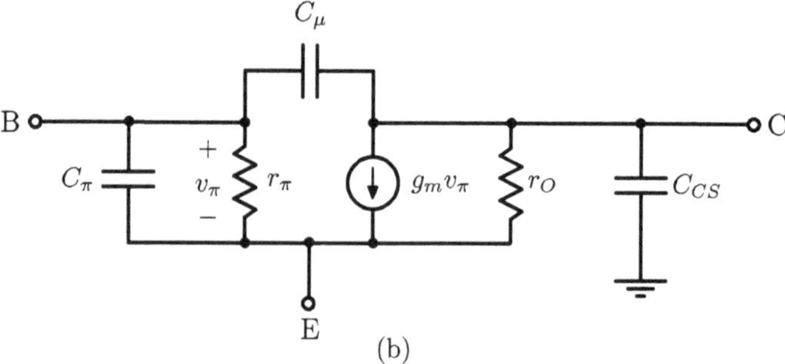

Fig. 3.1 **a** Integrated bipolar transistor structure and **b** small-signal equivalent circuit

The reverse-biased collector-substrate junction gives rise to a junction capacitance, C_{CS}. The three parasitic capacitance elements model a simplified frequency dependence of the bipolar device. Accurate determination of parasitic capacitances is required since these allow for f_T and f_{max} to be specified. The transit frequency is

$$f_T = \frac{g_m}{2\pi\left(C_\pi + C_\mu\right)} \tag{3.1}$$

and the maximum oscillation frequency is

$$f_{max} = \sqrt{\frac{f_T}{8\pi C_\mu r_b}} \tag{3.2}$$

where r_b (omitted from Fig. 3.1b) is the physical base resistance.

3.1.2 Integrated BJTs

The sequencing of doped silicon layers can be applied vertically (resulting in vertical BJTs) or horizontally (resulting in lateral BJTs). A basic integrated BJT consists of a heavily doped N+ emitter region, an N-type epitaxial layer, the P-type base area, and another heavily doped N+ buried collector region, all on top of a P-type substrate. P+ isolation areas isolate neighbouring devices on an IC. Advanced BJTs make use of trench isolation to increase density. The layout of a vertical NPN BJT is shown in Fig. 3.2 [2].

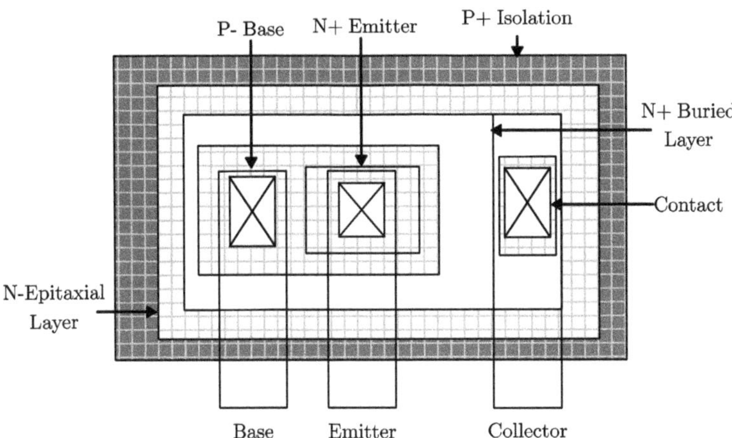

Fig. 3.2 Top view of a vertical PNP-BJT layout

The BJT structure contains several parasitic elements. The base resistance between the base contact to the active area, the collector resistance, resulting from the N-epitaxial layer and the emitter resistance, which is usually negligibly small. The junction between the N+ emitter and P base contains the emitter–base junction capacitance, C_{jE}. The N− collector area next to the base also gives rise to a collector–base junction capacitance C_{jC}.

3.1.3 Heterojunction Bipolar Transistors

Heterojunction Bipolar Transistors (HBTs) are the first silicon devices employing bandgap engineering. The SiGe HBT has evolved into a crucial device in RFICs. Figure 3.3 shows the cross-section of an HBT [3], followed by its small-signal forward-active equivalent circuit in Fig. 3.4 [4].

The process flow of an example 0.13 μm BiCMOS process by STMicroelectronics is compared to a standard digital CMOS flow and outlined in Fig. 3.5 [5].

The initial part of the process flow is used to form the HBT collector area with buried layer implantation, the deep-trench formation and collector epitaxy. Next, the shallow trench area is formed, and the N-well/P-well areas for the MOS transistors are implanted into the epitaxial layer. Thereafter gate and oxide layers are deposited thereafter. Fabrication of the SiGe HBT begins with opening the HBT area in the gate polysilicon, followed by implantation of the localised SiC collector. The emitter is etched into the extrinsic base oxide-poly-oxide-nitride stack, which is deposited earlier. Selective epitaxial growth of the SiGe:C base region is next performed, consisting of a cavity that is etched within the pedestal oxide. This provides self-alignment of the emitter and base structures. After D-shaped inside spacers are formed, the polysilicon emitter is deposited. Finally, the poly-emitter

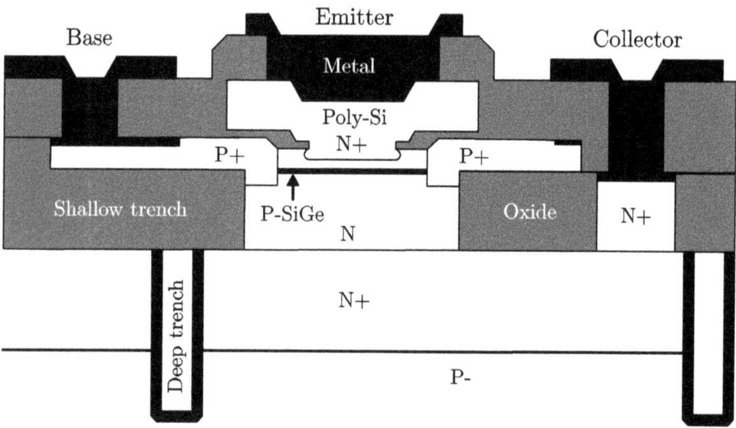

Fig. 3.3 Cross-section of a first-generation deep-trench SiGe HBT

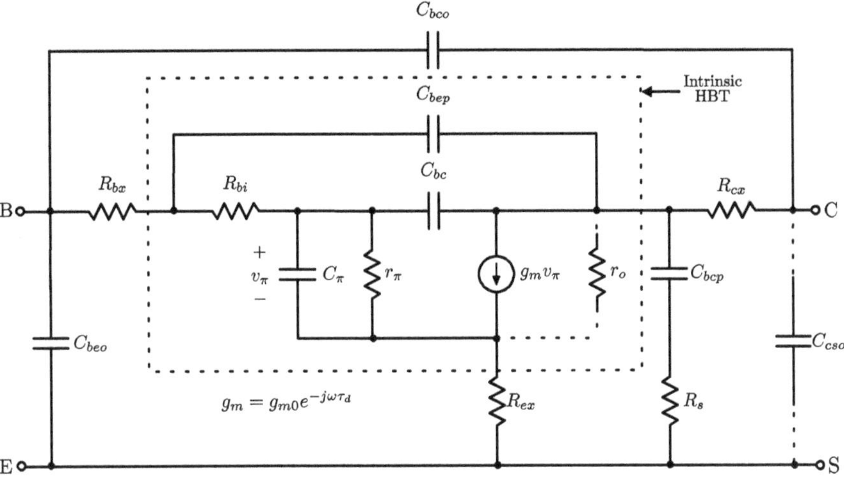

Fig. 3.4 Small-signal equivalent circuit of a deep-trench SiGe HBT

and extrinsic base polysilicon regions are patterned, which protects the MOS transistor gate during all the bipolar steps. In this process flow, the HBT is completed prior to forming the drain and source terminals of the MOS transistor to minimise the bipolar thermal budget impact.

3.2 MOS Transistors

MOSFETs consist of a channel region that separates highly doped drain and source regions. Complementary variations are available, depending on the material (P-type or N-type) used to form the terminals and substrate. Carrier mobility determines device speed. NMOS carriers are electrons, while PMOS carriers are holes. Naturally, the electron mobility in silicon is much higher (about three times higher) than hole mobility, resulting in the almost exclusive use of NMOS devices in RF and mm-wave designs.

3.2.1 Small-Signal Equivalent Circuit

Integrated MOSFETs contain three critical capacitances. The gate oxide capacitance (equal to WLC_{ox}) exists between the gate and the channel. Then, there is capacitance associated with the source-bulk and drain-bulk junctions, both in reverse bias. C_{ox} is inherently challenging to model because, technically, the model does not contain a

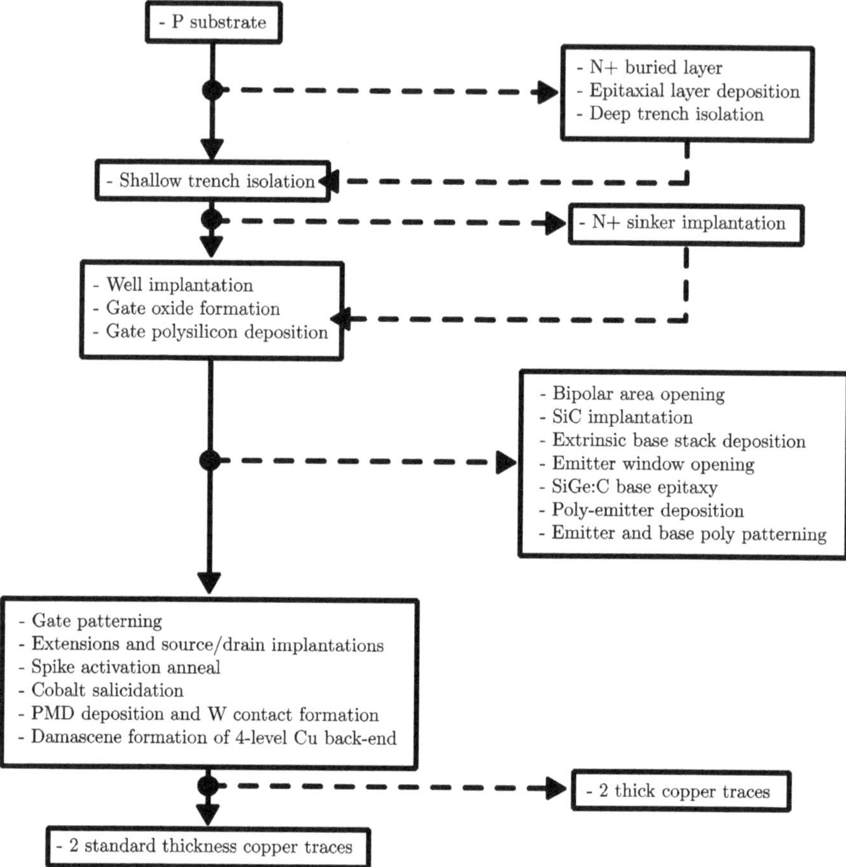

Fig. 3.5 0.13 μm BiCMOS process integration flow

channel [1]. This capacitance is therefore commonly separated, one between the gate-drain terminals and one between the gate-source terminals. The basic structure of a MOSFET is shown in Fig. 3.6a, followed by its small-signal equivalent in Fig. 3.6b.

3.2.2 Millimeter-Wave Operation

As the frequency of operation increases, additional resistive and capacitive components inherent to the fabrication process arise. The gate capacitance in a MOS device is arguably the most important parasitic component [6]. It is comprised of three distinct capacitances: the gate-source capacitance C_{GS}, the gate-drain capacitance

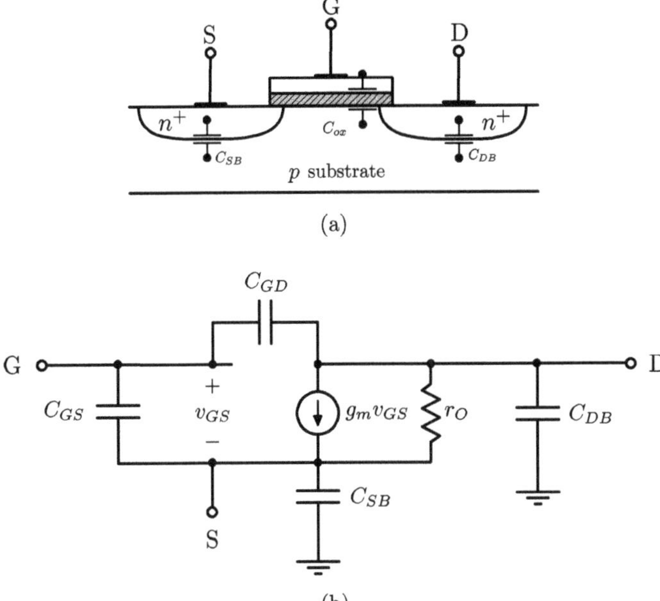

Fig. 3.6 **a** Structure and **b** small-signal integrated MOSFET transistor model

C_{GD} and the gate-bulk capacitance C_{GB}. Depending on the operating region of the device, C_G is divided differently. For example, in the saturation (forward-active) region,

$$C_{GS} = \frac{2}{3} C_{ox} W L \tag{3.3}$$

In the triode region, C_{GS} and C_{GD} are equal, and both are zero in cutoff. The source-body and drain-body capacitances only appear when their PN-junctions are in reverse bias as a result of the body effect. The small-signal equivalent circuit shown in Fig. 3.6b can be improved by taking the terminal resistances into account.

Additional contact resistances associated with the three terminals are also present and result from the physical material resistivity. These resistances are also bias-dependent, and even though they are typically small enough to be safely ignored, they can have a prominent effect if the device is operated close to its rated capability.

The transit frequency in a MOS device is given by

$$f_T = \frac{g_m}{2\pi \left(C_{gs} + C_{gd}\right)} \tag{3.4}$$

and the maximum oscillation frequency is

$$f_{max} = \sqrt{\frac{f_T}{8\pi C_{gd} r_g}} \tag{3.5}$$

where r_g denotes the parasitic gate resistance.

3.2.3 Layout Effects

Optimisation of MOSFET devices, particularly when operated at high frequencies where wide fingers are required, generally involves the use of multi-fingered devices. This reduces gate resistance, and from (3.5), increases f_{max}. Connecting the gate to both sides of the transistor fingers will further reduce resistance. Additionally, guard rings may be added to provide isolation from the substrate, which also serves to reduce noise. Section 3.3.2 covers the challenges of MOSFET modelling in greater detail.

3.3 Compact Modelling

Compact models link the design of a circuit to its fabrication. That is, devices fabricated on a wafer have their electrical properties represented by physics-based models that are sufficiently simple to support simulation and optimisation [7]. Device modelling is becoming increasingly important, given the immense rise in complexity, manufacturing cost and fabrication time associated with high-performance mm-wave devices. Throughout the industry, the first-pass success of high-frequency analogue circuits is highly sought after and, in most cases, a necessity to stay competitive.

3.3.1 HBT Models

For SiGe:C HBT technologies, several models of varying complexity are available. Perhaps the oldest compact model for bipolar transistors is known as Mextram, short for Most Exquisite Transistor Model. The high-current model (or HiCUM) is a more advanced physics-based model developed in the DOTSEVEN project [7]. The Vertical Bipolar Inter-Company (VBIC) Model improves the older SPICE Gummel-Poon (SGP) model, perhaps the most well-known bipolar model in existence.

3.3.1.1 Mextram

The Mextram model is an attempt at improving on the limitations of the classical Gummel–Poon bipolar model. It models the following parameters [8]:

- Charge storage effects
- High-injection effects
- Low-level, non-ideal base currents
- Hot-carrier effects in the collector epitaxial layer
- Effects of the parasitic PNP device and the substrate
- Intrinsic base effects
- Temperature dependence
- Quasi-saturation
- Weak avalanche breakdown
- Base region electric field
- Base–collector depletion capacitance

The Mextram model adds two nodes to its Gummel–Poon predecessor, which enables it to account for these additional effects. Moreover, about 35 of its 62 parameters must be determined by fitting to the transistor characteristics at a specific temperature. For the sake of simplicity, only the small-signal Mextram model will be shown here, in Fig. 3.7. The complete model can be viewed in the Mextram reference document [9].

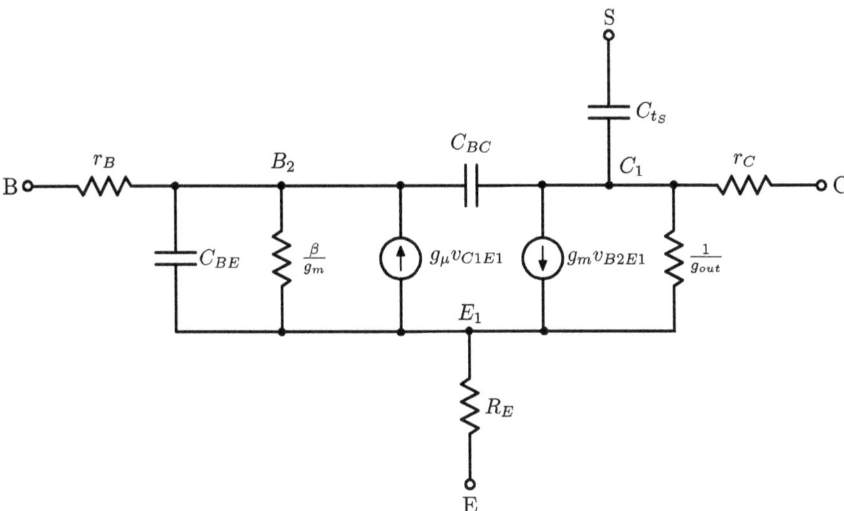

Fig. 3.7 Small-signal equivalent circuit of the Mextram model

3.3.1.2 HiCUM

The standard high-current model (HiCUM/L2) for bipolar transistors is widely used to address these requirements [7, 10, 11]. The HiCUM/L2 model is physics-based, whereby the value for each element in its equivalent circuit is a function of technology data (such as the collector doping beneath the emitter), electrical data (such as capacitance per unit area or sheet resistance), physical data (e.g. hole and electron mobility), transistor dimensions (number of contacts per terminal and emitter size), operating point and finally temperature. A single set of technical and electrical data can therefore be utilised to calculate an entire array of model parameters, enabling statistical modelling as well as circuit-level optimisation. An equivalent circuit diagram of the HiCUM/L2 model is shown in Fig. 3.8, and the networks used to model thermal effects and non-quasi-static (NQS) effects are shown in Figs. 3.8 and 3.9 [7].

The transfer current describes transistor behaviour as a current-controlled source. Firstly, the B^* node in Fig. 3.8, separating the operating point-dependent internal base resistance (R_{Bi}^*) from the operating point-independent external resistance (R_{B_x}), is indeed needed to account for emitter periphery effects, as these can have a substantial effect on modern transistor behaviour. The B^* node also aids in improving the modelling of the distributed nature of the external base–collector region, which can be done by separating the external base–collector capacitance (C_{BC_x}) over R_{B_x} in the shape of a π-network. The small-signal emitter current-crowding effect at high frequencies is also taken into account via the C_{rB_i} capacitance. More advanced technologies benefit from the emitter–base isolation capacitances ($C_{BE_{par}}$). Internal collector resistances are accounted for through a combination of the transfer current i_T and the minority charge, denoted by C_{dE} and C_{dC}. The internal collector terminal (denoted by C') is located physically on the edge of the epitaxial collector region.

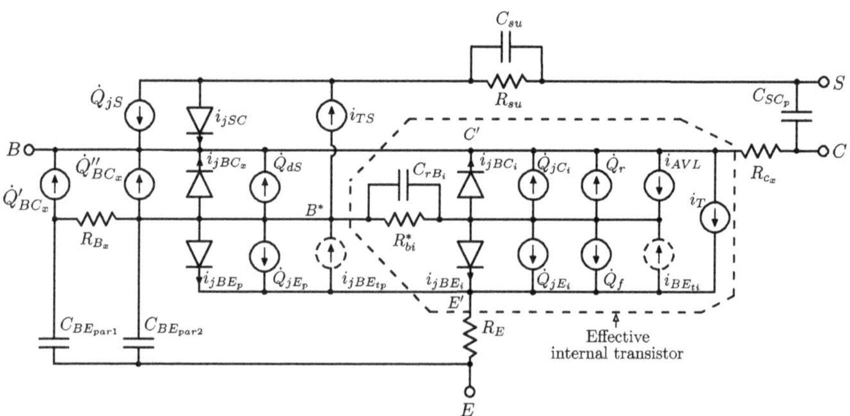

Fig. 3.8 HiCUM/L2 equivalent circuit diagram

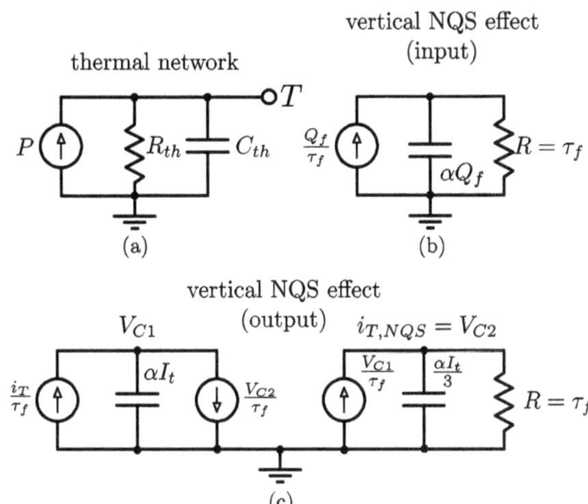

Fig. 3.9 HiCUM/L2 adjunct networks that model **a** electro-thermal effects, **b** input NQS effects, and **c** output NQS effects

In doing so, the model saves computational complexity, removes one node and is verified to be accurate for an extensive range of bipolar technologies.

3.3.1.3 VBIC

VBIC offers several modelling advantages over the older SGP model [12]:

- Improved early effect behaviour
- Quasi-saturation modelling
- Substrate transistor
- Parasitic oxide capacitance
- Avalanche multiplication
- Improved temperature dependence
- Decoupled base and collector currents
- Self-heating
- Improved HBT modelling

With additional effects turned off, the VBIC model closely approximates the behaviour predicted by the SGP model, save for the early effect. Figure 3.10 shows the VBIC equivalent circuit [4, 12]. The circuit includes a parasitic PNP transistor, an intrinsic NPN transistor and some parasitic resistances and capacitances.

While the equivalent circuit in Fig. 3.10 contains parasitic resistances and capacitances and some current sources, these should be treated as voltage-controlled and charge-controlled elements. Figure 3.11 shows the remaining thermal and excess phase networks of the VBIC model.

In most modern integrated designs, VBIC has replaced the SGP model as an industry-standard [12]. Generally, use of the VBIC model is recommended for higher density designs, while HiCUM performs better for high-current applications such as

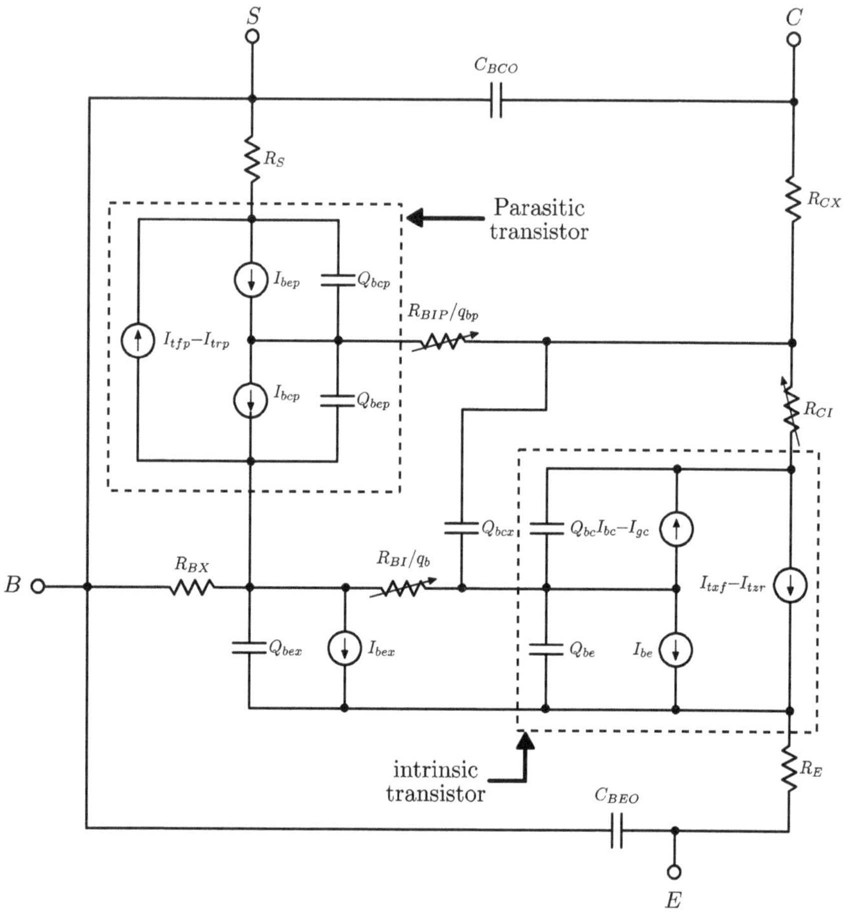

Fig. 3.10 VBIC transistor model equivalent circuit

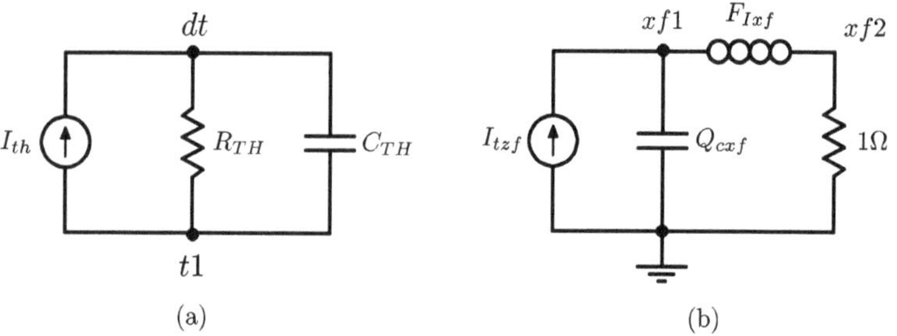

Fig. 3.11 Additional VBIC equivalent circuits **a** thermal network and **b** excess phase network

power amplifiers. The main reasons for the VBIC model performing poorly at high currents are summarised here [13]:

- Lightly doped collector areas in GaAs HBTs place higher importance on charge modulation of the parasitic collector capacitance
- HBTs have fairly high C_{BC} close to its quasi-saturation region of operation, which the VBIC model does not predict with sufficient accuracy
- Modelling of the Kirk effect requires that the transit time and the parasitic collector capacitance (which is dependent on current) are modelled with greater accuracy than what the stock VBIC model can achieve
- Accurate modelling of the reverse injection of minority carriers between the base and emitter is essential at higher current levels
- Reverse transit time is dependent on junction currents

However, some efforts have been made to improve the high-current behaviour of the standard VBIC model [13]. Most approaches focus on improving the modelling of the intrinsic device.

3.3.2 FET Models

MOS device modelling is a different story. The later sections of this chapter will discuss the complexities associated with MOS modelling, as there are numerous physical effects (many of them layout-dependent) that can introduce severe inaccuracies if not adequately accounted for. MOS models developed by the BSIM group are most widely used throughout the industry.

Models for MOS transistors started appearing over 50 years ago [14]. The reference approach—the Pao-Sah model—was one of the widely used models early on. However, its use of double integrals made it impractical for simulators at the time. The Shichman–Hodges model was the first iteration of SPICE models for MOS transistors, which was nearly symmetric. About a decade later, the first practical models began to appear, based on surface potential ψ_s. Unfortunately, this approach required many years of development to solve difficulties with its implementation, before it became usable. As a consequence, MOS model development followed the route of source-referenced, threshold voltage-based models. Problems with these models have been known for at least 30 years, and hence development stagnated. Over time, as benchmarks were defined and eventually turned into requirements for model performance, the source-referenced, voltage threshold-based approach was retired. Modern MOS models are typically based on inversion charge, such as the BSIM-BULK model, or ψ_s-based, as used in FinFET, UTTB and bulk structures.

3.3.2.1 Challenges

Modelling of MOS devices by themselves is not complex since their principle of operation is relatively simplistic. However, immense complexity arises from the interaction between the MOS device and its surroundings. McAndrew summarises the proximity effects that complicate MOS transistor layout, which includes at least twenty different phenomena to take into account [14].

Transistor structures typically used to characterise layout-dependent properties use layouts with consistent spacing to strain sources. Fortunately, this method works reasonably well for irregular layouts, but characterising these properties is still challenging. For example, when trying to characterise layout effects that require seven geometries alongside three aggressor edge spacings (each with their own layout-dependent effects), a total of 105 test structures per device will result [14] if five types of proximity effects are to be included in the process. Taking into account the interactions between layout-dependent effects in this example would result in 1701 test structures, which is wholly impractical. In addition to this, shrinking feature sizes (e.g. with FinFETs) correspond to more significant contributions of parasitic resistors and capacitors throughout the device [15].

3.3.2.2 Bsim

The BSIM group began MOS transistor modelling in the 1980s. In 1996, the Compact Model Council selected BSIM3 as the industry-standard MOSFET model and subsequently, this was replaced by BSIM4 in 2000 [16]. BSIM4 is very flexible, and extracted data from technologies ranging from simpler 0.35 μm all the way to 28 nm can be fit with relative ease. BSIM3 and BSIM4 are threshold-based models, and they consist of a core model that represents about 20% of the model's behaviour via a simple long-channel MOSFET [17]. The remaining 80% consists of multiple models used to describe the effects of the actual device, which contributes primarily to the model's accuracy. Without these additional models, inaccuracies associated with device sizing could potentially be massive.

BSIM introduced novel models for short-channel effects, quantum-mechanical effects, non-uniform doping effects, $1/f$ noise, the strain effect, intrinsic input resistance, inter-band tunnelling and output conductance [17, 18]. The accuracy and simulation speed of the BSIM family of models is the main reason behind its widespread adoption throughout the industry. Analog and RF designs are plagued by a seemingly insignificant asymmetry around $V_{ds} = 0$ that originates mainly from BSIM's threshold-based core and its implementation of bias-dependent properties. As such, the BSIM group began with BSIM6 development in 2010, aimed at implementing a new core model that addresses the asymmetry problems. By 2013, BSIM6 had been declared the new industry standard.

3.4 Process Corner Modelling

Modern semiconductor processes push transistor dimensions to their limits, and as a result, process development is becoming exponentially more difficult. The improvements in power, density and performance introduced by semiconductor ICs with each new generation thus place tremendous pressure on accurate device modelling. Simulation methodologies and device modelling must therefore be improved upon simultaneously, so that circuit predictions can be more accurate in the presence of process variations [19].

Process variations are categorised as within-die and die-to-die. The stochastic nature of semiconductor fabrication results in unavoidable variance. On the other hand, variation between dies results from restrictions with process control, such as layer thickness tolerances between wafers. Nonetheless, any process variation affects circuit performance, and the performance of *all* fabricated ICs must be adequate. To achieve this, accurate determination of process corners is necessary.

3.5 Conclusion

This chapter covered the development and challenges associated with highly scaled semiconductor silicon technologies. A strong focus on compact models stems from the growing demand for the first-pass success of mm-wave designs, an absolute necessity for remaining competitive in today's market.

Most SiGe process design kits (PDKs) provide VBIC and HiCUM models for their bipolar devices, providing the designer with additional flexibility depending on the intended application. For example, the HiCUM model is well suited for high-current applications such as PAs, where self-heating can be a significant issue, and accurate modelling is highly beneficial.

MOSFET modelling is arguably in a better state compared to HBT. Short-channel Insulated-Gate-Field-Effect-Transistor (IGFET) models (such as the ones used in BSIM models) have been used for a wide range of operating frequencies and tremendous effort has gone into maintaining these models to keep up with technology scaling.

References

1. Razavi B (2014) Fundamentals of microelectronics, 2nd edn. John Wiley & Sons Inc., Hoboken, New Jersey
2. Saha SK (2016) Compact models for integrated circuit design: conventional transistors & beyond. CRC Press, Boca Raton, Florida
3. Zhu C et al (2005) Damage mechanisms in impact-ionisation-induced mixed-mode reliability degradation of SiGe HBTs. IEEE Trans Device Mater Reliab 5(1):142–148. https://doi.org/10.1109/TDMR.2005.843835

4. Lee K et al (2005) Direct parameter extraction of SiGe HBTs for the VBIC bipolar compact model. IEEE Trans Electron Devices 52(3):375–384
5. Avenier G et al (2009) 0.13 μ m SiGe BiCMOS technology fully dedicated to mm-wave applications. IEEE J Solid-State Circuits 44(9):2312–2321. https://doi.org/10.1109/JSSC.2009.2024102
6. Božanić M, Sinha S (2020) Millimeter-wave integrated circuits. Springer International Publishing, New York City, New York
7. Rinaldi N, Schröter M (eds) (2018) Silicon-germanium heterojunction bipolar transistors for mm-Wave systems: technology, modeling and circuit applications. River Publishers, Gistrup, Denmark
8. Kloosterman WJ, Geelen JAM, Klaassen DBM (1995) Efficient parameter extraction for the MEXTRAM model. In: Proceedings of the IEEE bipolar/BiCMOS circuits and technology meeting, pp 70–73. https://doi.org/10.1109/bipol.1995.493869
9. van der Toorn R, Paasschens JCJ, Kloosterman WJ (2008) The Mextram bipolar transistor model: level 504.7. Delft University of Technology
10. Schröter M, Pawlak A (2017) HiCUM/L2 A geometry scalable physics-based compact bipolar transistor model. Available: https://www.iee.et.tu-dresden.de/iee/eb/forsch/Hicum_PD/Hicum23/hicum_L2V2p4p0_manual.pdf. Accessed 14-Mar-2021.
11. Schröter M, Chakravorty A (2010) Compact hierarchical bipolar transistor modeling with HiCUM. World Scientific Publishing Company
12. Cao X, McMacken J, Stiles K, Layman P, Liou JJ, Ortiz-Conde A (2000) Comparison of the new VBIC and conventional Gummel–Poon bipolar transistor models. IEEE Trans Electron Devices 47(2):427–433. https://doi.org/10.1109/16.822290
13. Wei CJ, Gering JM, Tkachenko YA (2005) Enhanced high-current VBIC model. IEEE Transactions on Microwave Theory and Techniques 53(4 I):1235–1243. https://doi.org/10.1109/TMTT.2005.845715
14. McAndrew CC (2019) Compact models for MOS transistors: successes and challenges. IEEE Trans Electron Devices 66(1):12–18. https://doi.org/10.1109/TED.2018.2849943
15. Hisamoto D et al (2000) FinFET-A self-aligned double-gate MOSFET scalable to 20 nm. IEEE Trans Electron Devices 47(12):2320–2325. https://doi.org/10.1109/16.887014
16. Chauhan YS et al (2012) BSIM industry standard compact MOSFET models. In: European solid-state circuits conference, pp 30–33. https://doi.org/10.1109/ESSCIRC.2012.6341249
17. Chauhan YS et al (2014) BSIM6: analog and RF compact model for bulk MOSFET. IEEE Trans Electron Devices 61(2):234–244. https://doi.org/10.1109/TED.2013.2283084
18. Dunga MV, Lin CH, Xi X, Lu DD, Niknejad AM, Hu C (2006) Modeling advanced FET technology in a compact model. IEEE Trans Electron Devices 53(9):1971–1977. https://doi.org/10.1109/TED.2005.881001
19. Dongaonkar S, Mudanai SP, Giles MD (2019) From process corners to statistical circuit design methodology: opportunities and challenges. IEEE Trans Electron Devices 66(1):19–27. https://doi.org/10.1109/TED.2018.2860929

Chapter 4
Passive Circuits and Building Blocks in Millimeter-Wave Silicon

Silicon mm-wave circuits play a crucial role in many future wireless systems. Satellite communications, 5G new radio, automotive radar and high-resolution imaging are some applications that rely on mm-wave silicon. While low RF operation permits lumped element designs for passive circuits, mm-wave circuits rely on distributed and hybrid elements to alleviate the difficulties associated with silicon substrates at such high frequencies. Primarily, substrate losses and constrained chip area contribute to the complexity of implementing high-performance passive circuits. Passive building blocks such as couplers, baluns, combiners, filters and resonators are used extensively in mm-wave ICs. The literature on passive components and building blocks is mainly concerned with miniaturisation while maintaining adequate performance. Chip area is costly, and passive circuits consume plenty of it relative to active circuits.

Passive circuits have a marked effect on surrounding circuits. Frequency multipliers, for example, require high-performance baluns with excellent phase and gain error magnitude or are doomed to suffer from poor fundamental suppression [1, 2]. In mm-wave amplifiers, power combiners play a pivotal role in the power-added efficiency since combiner loss and area efficiency have a direct impact [3].

On-chip silicon passive circuits offer greater flexibility compared to traditional planar circuits. The main reason for this is the multiple thick metal layers that are available in most modern technologies, which, in turn, enable broadside coupling techniques. One example of this is broadside-coupled meander lines that can be used to implement resonators with tiny footprints [4, 5, 6, 7]. Folded topologies have also been proposed, where resonators are created and laid out over multiple metal layers, significantly reducing chip area [8]. Tuning of the self-resonant frequency of on-chip inductors is another method used to achieve miniaturisation. Multiple turns or 3D folded structures can be used to control self-resonance [9].

RF CMOS and SiGe BiCMOS design kits offer similar sets of passive components. Bipolar process steps complement the front end-of-line (FEOL) of the process, enabling PIN and Schottky diodes [10]. Moreover, standard BiCMOS technologies offer mm-wave back end-of-line (BEOL), while this is not standard for most RF

J. du Preez and S. Sinha, *State-of-the-Art of Millimeter-Wave Silicon Technology*,
Lecture Notes in Electrical Engineering 945,
https://doi.org/10.1007/978-3-031-14655-8_4

CMOS technologies. Additional thick metallisation layers provide better current handling capability for high power applications and enable higher Q-factor inductors, improving filter and transformer performance.

4.1 Matching Circuits and Impedance Transformation

Active devices rarely present purely resistive impedances at their input and output ports. Impedance matching is used to transform between varying impedance values—whether it is to match an antenna port to the input of a low-noise amplifier or the output of a voltage-controlled oscillator to a mixer input. To achieve maximum power transfer, conjugate matching must be employed, which means that the load impedance is the conjugate of the source impedance, $Z_L = Z_S^* = R - jX$. This essentially means that the phase difference is zero and that the source and load see identical, real impedances. However, this condition is predicated on the assumption that the passive components (transmission lines, inductors, etc.) used to implement the matching networks are lossless. The design becomes increasingly complex when non-ideal passive components are used.

Other than power transfer, other performance metrics are also affected by impedance matching. Active devices such as transistors may be matched to two additional impedance targets to achieve different design goals:

- Z_{opt} represents the input impedance that results in the lowest noise figure.
- Z_{gain} represents the input impedance that achieves maximum gain.

Practical active devices have feedback between the output and input ports (that is, $S_{12} \neq 0$), which means that the impedance target for maximum gain differs slightly from S_{11}^* [11].

4.1.1 Matching Network Losses

Figure 4.1 shows an impedance matching network designed to maximise the efficiency of power transfer between the source (V_s) and load (Z_L).

The situation depicted in Fig. 4.1 is all too familiar. The goal is to achieve maximum power transfer from the source with impedance $Z_s = R_s(1 - jQ_s)$ through a two-port network to a load, R_L. Non-ideal effects in the source are accounted for by considering its resistance R_s and impedance quality factor Q_s in the frequency band of operation. In an ideal case, the matching network transforms R_L into Z_{in}^*. Since the lossless network dissipates no power, $R_L = Z_{in}^*$.

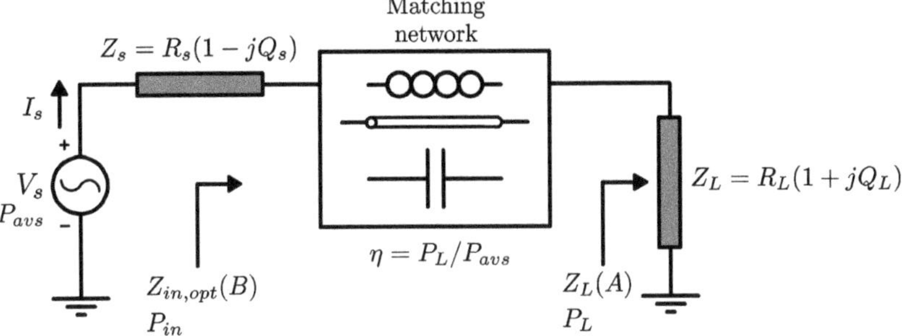

Fig. 4.1 Impedance matching network

4.2 Power Combiners and Directional Couplers

Directional couplers and power dividers/combiners are passive circuits used for power combining and splitting. A divider or splitter separates an input signal into two or more output signals with lower power, and the opposite is valid for a combiner. Couplers and dividers typically have three or four ports but in some cases can have more. Three-port networks are usually structured as T-junctions, while four-port networks are hybrids and directional couplers. Power splitters may provide in-phase or out-of-phase output signals, in addition to equal or unequal power division. Hybrids typically have equal power distribution, while directional couplers are designed with arbitrary power division.

Several properties must be accounted for in evaluating power splitters and combiners. Combiner loss is of particular importance, especially after the final gain stage in an amplifier design, since it affects efficiency. Combiner size must be minimised, given that IC cost is proportional to its size. This is especially challenging in large power combining amplifier topologies [12]. However, some combiners may require lengthy lines for phase matching purposes. Combiner bandwidth is essential and problematic in mm-wave designs where large absolute bandwidths in the GHz range are required. Power amplifiers (PAs) with large power combining schemes can benefit significantly from combiner networks that feed DC bias to the active devices. Lastly, larger structures that approach the signal wavelength in dimensions can have a negative effect on odd-mode stability.

4.2.1 Power Combining Performance Metrics

On-chip power combining techniques are compared based on two important figures of merit: area efficiency and spatial power density [13]. Compounding losses in M-way combiner networks (exacerbated by the requirement for large networks at the input and output of the PA) necessitate maximising area efficiency. As splitter networks grow, so too must the combiner network, and adding more devices to the combining scheme results in an exponential area increase. Furthermore, the loss associated with large combiner networks means that the output power eventually saturates. The area efficiency is

$$\eta_{\text{area}} = \frac{\text{Active area (splitter + combiner)}}{\text{Number of combined PA cells}}\,\text{mm}^2 \quad (4.1)$$

The thermal limitations of the technology (primarily, the thickness of metal layers, corresponding to current density limits) used necessitate computing the spatial power density, which is

$$\text{Spatial power density} = \frac{P_{out}}{\text{Active area (splitter + combiner)}}\,\text{W/mm}^2 \quad (4.2)$$

4.2.2 *Mm-Wave Combiner Challenges*

Power combiners can either be matched (such as with 90° couplers and Wilkinson combiners) or not (as is the case with transformer combiners). This presents a fundamental tradeoff in combining at mm-wave frequencies. The combiner that is 50Ω-matched provides a more straightforward implementation, but the networks needed to achieve 50Ω matching consume chip area. Conversely, combiners not matched to 50Ω achieve power combining and impedance transformation while consuming little area. These combiners are typically more challenging to design and result in complex splitter networks.

4.2.2.1 *Issues with Impedance Transformation*

Monolithic transformers are often used in power combining circuits. They allow, firstly, transfer of the output voltage and impedance to the secondary windings and, subsequently, are stacked to match the output load impedance. For mm-wave PAs, transistor sizing plays an essential role in the design [14, 15, 16]. Larger transistors that can handle higher current densities have reduced gain because of increased parasitic capacitances [3]. The design requirements that simultaneously drive the transistor size and the transformer dimensions are thus contradictory, which severely limits the impedance freedom that the transformer can provide. Additionally, the current density limitation attached to the metal layers of a typical CMOS or BiCMOS process places a hard limit on the DC feed design [17, 18].

4.2.2.2 Chip Area Limitations

Combiner networks used in PA design generally require full integration since this minimises chip area and simplifies fabrication. Furthermore, integration into phased array transmitters is more accessible when the PA portion occupies less area [19]. The planar layout of transformer combiners is especially challenging and explodes in complexity when more than two PA cells are combined. 3D architectures have been suggested to tackle this problem [20].

4.2.2.3 Symmetry Requirements

The reduced wavelength in mm-wave designs means that combiner networks are much more sensitive to asymmetry. As the number of combiner and splitter branches increases, it becomes increasingly challenging to phase match each branch. In mm-wave PAs, power combining is imperative. While conventional planar combiner networks are easy to route symmetrically, the network losses and limitations with current density limit the number of PAs that can be efficiently combined. On the other hand, transformer combiners present a formidable challenge in phase matching combiner branches, and the optimal approach is therefore dependent on the application requirements.

4.2.2.4 Instabilities

Mm-wave power combined amplifiers can suffer from instability. Larger devices are often used to improve the output power capabilities at the cost of larger parasitic capacitance (specifically the collector–base capacitance in SiGe HBTs or the gate-drain capacitance in MOSFETs) [21]. Moreover, the increase in parasitic capacitance leads to lower reverse isolation (S_{12}), potentially leading to the generation of sub-harmonics and spurious oscillations. With SiGe HBTs, this problem is worse because of the voltage dependence of C_{CB} and the large voltage swings present in Class E PAs [22]. Combining multiple PAs in parallel can also excite even- and odd-mode oscillating behaviour in the internal nodes of the switching devices. Using half-harmonic RC traps in conjunction with Wilkinson combiners can prevent even- and odd-mode oscillations, simply because the common-mode resistance between the combiner branches guarantees that the loop gain of the RC trap is below 1 [23].

4.2.3 T-Junction Dividers

The simplest power dividers are T-junctions—three-port networks that may be implemented in just about any kind of transmission line topology. Figure 4.2 shows a transmission line model of such a divider [24].

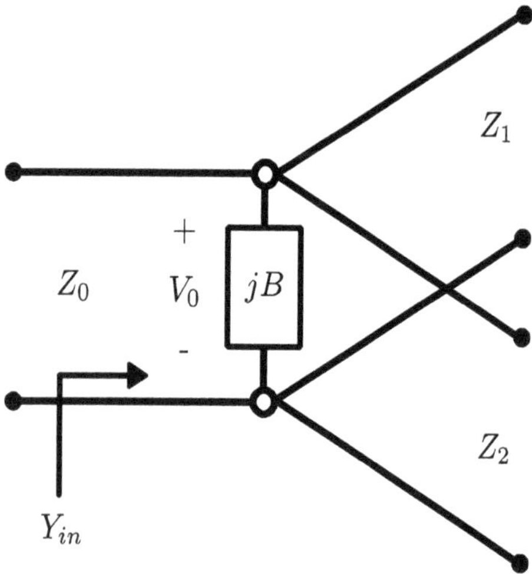

Fig. 4.2 Transmission line equivalent of the T-junction splitter

The model in Fig. 4.2 describes just about any lossless T-junction divider. Higher-order modes and fringing fields that arise from the discontinuities in the junction are accounted for with a shunt susceptance (B). Practical T-junction splitters implement tuning elements to cancel out B over a limited bandwidth. Additionally, Z_1 and Z_2 are tuned to create the desired power division ratio. For example, when $Z_0 = 50\Omega$, select $Z_1 = Z_2 = 100\Omega$ to obtain an equal 3 dB split. The T-junction divider does not provide any isolation between the two output ports, which means that the impedance looking into either output will not be matched. However, by adding lossy components such as series resistors to each path (with impedance $Z_0/3$) matching can be achieved at all ports. These are known as resistive dividers.

4.2.4 *Wilkinson Dividers*

Neither T-junction dividers nor resistive dividers provide isolation between output ports. The Wilkinson divider improves both the topologies mentioned earlier, since it appears lossless when the outputs are matched (Fig. 4.3).

The Wilkinson splitter can be implemented with microstrip or striplines, but coplanar waveguide (CPW) lines are more common in mm-wave designs. Distributed Wilkinson dividers use quarter-wave transmission lines and thin-film resistors on an appropriate substrate. They induce low losses and are therefore suitable for use as the final combiner. Furthermore, they provide sound isolation, which means that chip degradation will occur gradually in the event of a failure [25]. In-phase combining of larger devices is possible because of the length of quarter-wave microstrip sections, and typical process variations do not significantly affect performance.

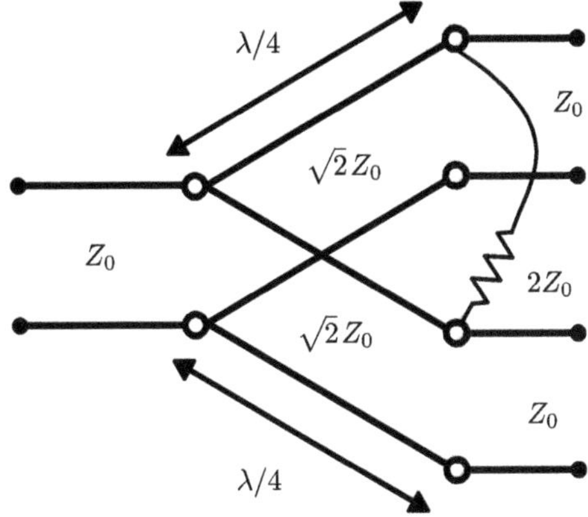

Fig. 4.3 Transmission line equivalent of a Wilkinson divider

4.2.5 *Capacitive Combiners*

Figure 4.4a shows a cascaded two-way capacitive combiner implementation, and Fig. 4.4b shows a direct four-way variation

Capacitive power combiners offer some advantages over traditional quarter-wavelength resistive combiners. Transmission lines that are shorter and wider may be used, leading to lower combiner loss [26]. However, it can lead to some degree of bandwidth degradation, which is to be avoided. The theoretical bandwidth limit can be determined using the Bode-Fano criteria,

$$\int_0^\infty \ln \frac{1}{|\Gamma(\omega)|} d\omega \leq \frac{\pi}{R_L C_L} \tag{4.3}$$

where R_L and C_L are the equivalent load seen by the output of the combiner. Assuming a modest reflection coefficient of -20 dB (or $\Gamma = 0.1$) at the design frequency, and using typical values for a mm-wave power amplifier (PA) load ($R_L = 35\Omega$ and $C_L = 80$ fF) leads to

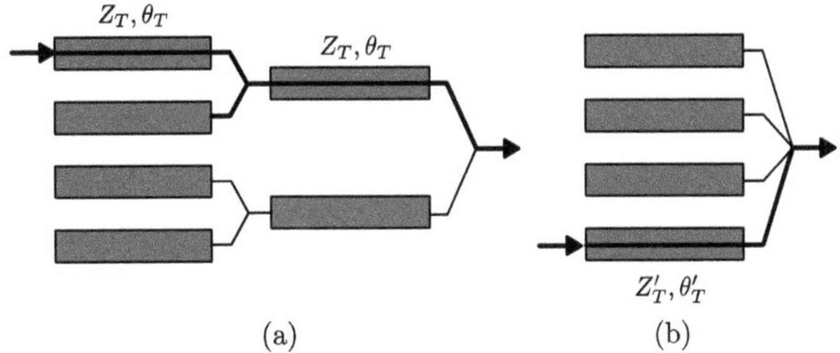

Fig. 4.4 **a** Cascaded two-way and **b** direct four-way capacitive combiners

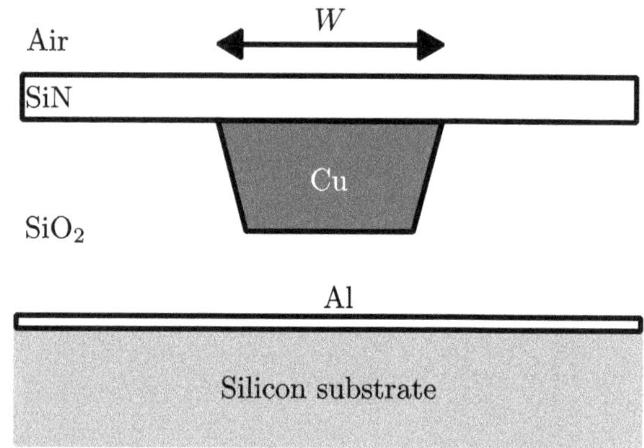

Fig. 4.5 Transmission line cross-section for the capacitive combiner branches

$$2\pi \ln \Delta f \cdot \ln \frac{1}{0.1} \leq \frac{\pi}{35 \cdot 80 \times 10^{-15}} \tag{4.4}$$

This, in turn, results in a possible 77.5 GHz of impedance transformation bandwidth, exceeding any realistic requirement in the mm-wave range. DC operation also benefits from short, low-impedance microstrip traces since it improves the feeding of bias voltages to the PAs. Power loss is strongly dependent on line length, so it is only logical that the shortest possible traces must be used. Figure 4.5 shows the cross-section of a transmission line that can be used for such combiners in 65 nm CMOS [26].

Losses affecting the transmission line in Fig. 4.5 are radiation, dielectric and conductive loss. Radiation loss in this configuration is negligible, while the metal conductive loss is reasonably large and must be considered. Increasing the trace width (W in Fig. 4.5) reduces top metal loss, but it increases bottom metal conductive loss due to the increase in current density flowing in the bottom metal layer. The power loss for the circuit in Fig. 4.4a along the $2\theta_T$ line is

$$P_L = 20 \log\left(e^{\alpha\lambda 2\theta_T/2\pi}\right) \tag{4.5}$$

The dependence on line length ($2\theta_T$) is observed in (4.5).

4.2.6 Combiner Applications

A practical application of Wilkinson combiners is in PA circuits [13, 27]. Figure 4.6 shows the use of six identical Wilkinson splitters used to feed and subsequently combine four PA cells. Note that the resistors are omitted from Fig. 4.6 for the sake of simplicity.

Datta et al. report a Q-band Class E PA using a two-way Wilkinson combiner/divider scheme implemented in a 0.13 μm SiGe BiCMOS process

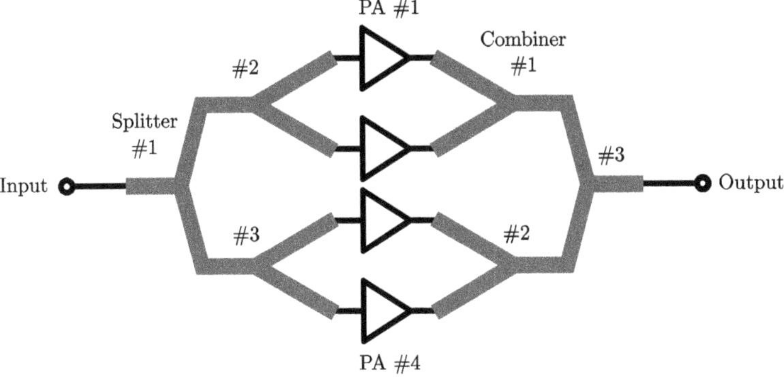

Fig. 4.6 Wilkinson combiner used in a 4-way PA scheme

Fig. 4.7 Transformer-based voltage-combining amplifier

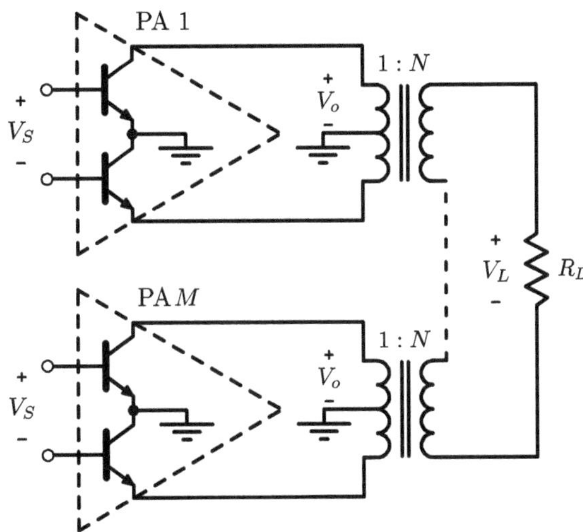

(IBM8HP) [28]. The high-efficiency Class E design places additional strain on the combiner network, requiring exceptionally low loss. The main challenge at Q-band frequencies in silicon design is to implement low-loss transmission lines with higher impedance ($Z_0\sqrt{2}$) with the on-chip metal layers.

Figure 4.7 shows a voltage-combining transformer scheme used to realise power combining and impedance transformation simultaneously.

4.2.7 *Quadrature Hybrid Couplers*

Quadrature hybrids are 3 dB directional couplers where the through and coupled ports have a 90° phase offset. Quadrature hybrids are applied widely in microwave and mm-wave circuits, such as frequency triplers, phase shifters and I/Q modulators and demodulators.

4.2.7.1 Branch-Line Coupler and Coupled Lines

Planar quadrature hybrids implemented with microstrip or stripline are called branch-line couplers, and Fig. 4.8 shows the layout of one such coupler.

Microstrip lines implemented in silicon substrates are plagued by low effective dielectric constant (ε_{eff}) due to the surrounding silicon dioxide. This results in poor area efficiency. Figure 4.9 shows an improvement of the branch-line coupler suitable for mm-wave designs, implemented with shielded slow-wave lines [29, 30]. Unlike traditional shielded slow-wave designs, the one shown in Fig. 4.9 does not make use of side ground planes.

The slow-wave transmission lines use quarter-wave sections for the lower ($Z_0/\sqrt{2}$) and higher (Z_0) impedance branches. Using slow-wave lines results in shorter wavelengths and a lower net phase velocity, which means that this approach can achieve better area efficiency with lower signal loss. This implementation provides a branch-line coupler that is roughly 75% smaller than the traditional design (Fig. 4.8).

Unshielded transmission lines routed close to each other will inevitably couple due to electromagnetic field interaction between the two lines. Coupled transmission lines consist of three conductors, although more may be used.

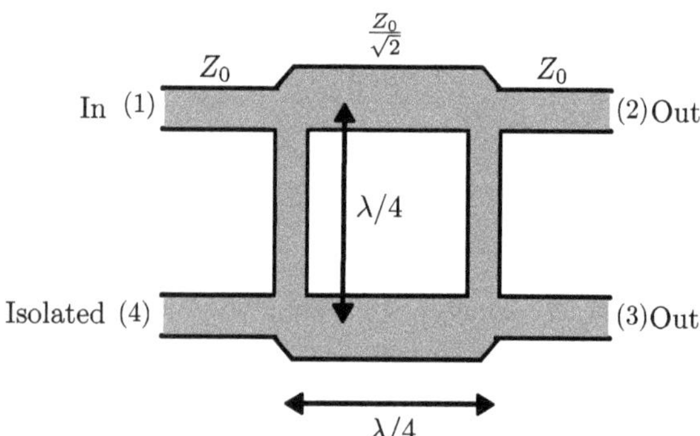

Fig. 4.8 Planar layout of a branch-line coupler

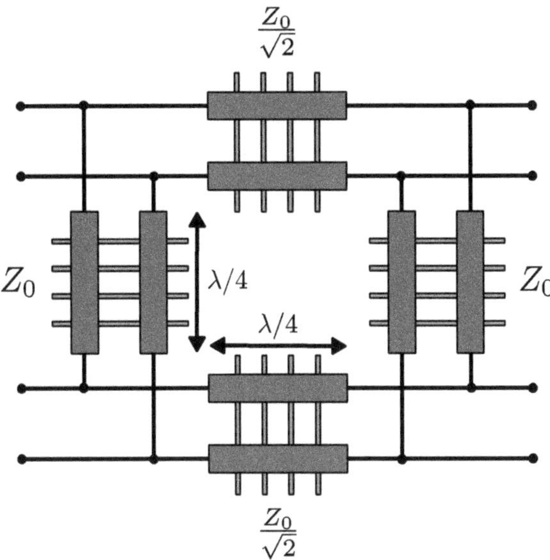

Fig. 4.9 Differential shielded slow-wave branch-line coupler

4.2.7.2 Lange Coupler

Coupling between coupled lines is often not strong enough to achieve 3 dB or 6 dB coupling factors. Coupling strength between edge-coupled lines can be improved by placing multiple lines in parallel, and Lange couplers are a popular implementation of this concept. Lange couplers are also 3 dB, 90° circuits where four coupled lines are placed in parallel along with interconnects. Figure 4.10 shows a microstrip Lange coupler [24].

Lange couplers can easily achieve 3 dB coupling, with bandwidths of at least an octave. The two coupled and through ports (ports 2 and 3 in Fig. 4.10) have a 90°

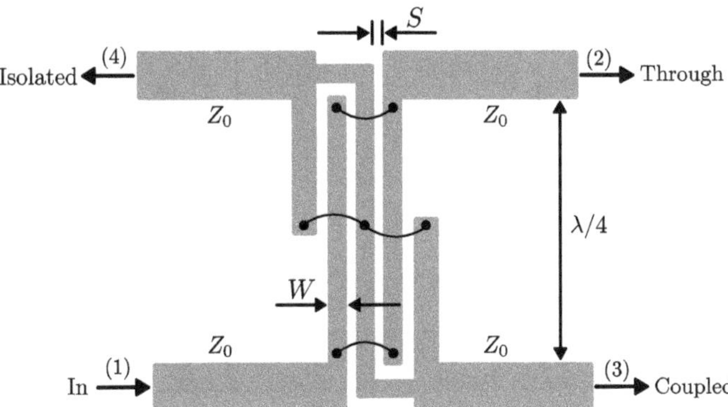

Fig. 4.10 Microstrip Lange coupler

phase difference between them. However, Lange couplers occupy a relatively large chip area and only provide two-way power splitting.

4.2.7.3 Slow-Wave Coplanar Couplers

The coupled slow-wave CPW (CS-CPW) efficiently implements coupled lines in silicon substrate [31]. It consists of a pair of signal lines surrounded by ground planes and a grid-like metal shield below the signal-ground pairing. Figure 4.11 shows a cross-sectional view of a CS-CPW.

The ground planes routed next to the signal traces create a strong ground return path for even-mode propagation. Electromagnetically speaking, this is a fairly complex structure and computationally intensive to simulate. Efficient modelling is thus critical for a successful design [30, 32].

4.2.8 Coupler Applications

4.2.8.1 Power Detectors

On-chip power sensing is reliant on couplers. Modern integrated transceivers make use of power sensing to estimate load power as well as gain [33]. Figure 4.12 shows a coupler-based power detection scheme connected to the load output of an integrated PA.

The coupling factor is kept to a minimum since it is undesirable to disturb the power flow out of the PA, and reflections due to mismatch must be avoided. The load

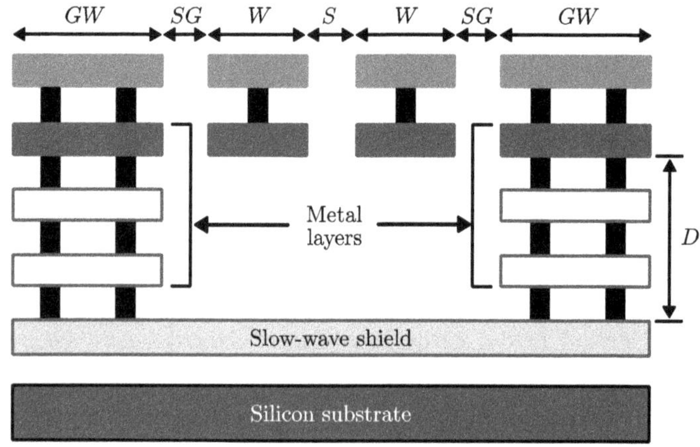

Fig. 4.11 Cross-section of a CS-CPW

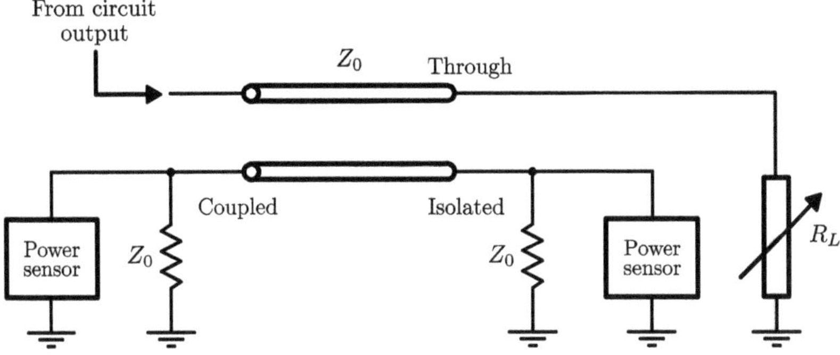

Fig. 4.12 Practical power detector application

in Fig. 4.12 is shown as variable since slight mismatches earlier on as well as temperature fluctuations mean that R_L will often deviate from the ideal 50Ω. Lower RF and perhaps even microwave power detection circuits may be implemented with transformers, but mm-wave designs use coupled transmission lines almost exclusively [34].

4.2.8.2 Doherty Amplifiers

Many applications require high linearity, especially ones that utilise modulation schemes with high peak-to-average power ratio (PAPR), such as quadrature amplitude modulation (QAM). In such cases, intermodulation distortion and harmonics resulting from poor linearity will destroy the signal quality and render the demodulation process fruitless. The higher conduction angle associated with linear PAs degrades their efficiency since voltage and current waveforms across the amplifying transistor overlap, causing power dissipation in the device [35].

The Doherty amplifier is essentially an efficiency enhancement technique, where the primary PA is used in conjunction with an auxiliary amplifier. The auxiliary PA will switch off when the input drive is sufficiently low, resulting in significant efficiency improvements at backed-off power. Figure 4.13 shows a classical Doherty PA.

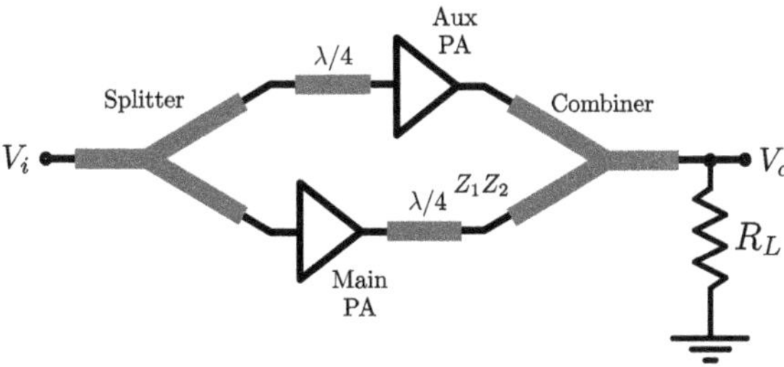

Fig. 4.13 Doherty amplifier schematic

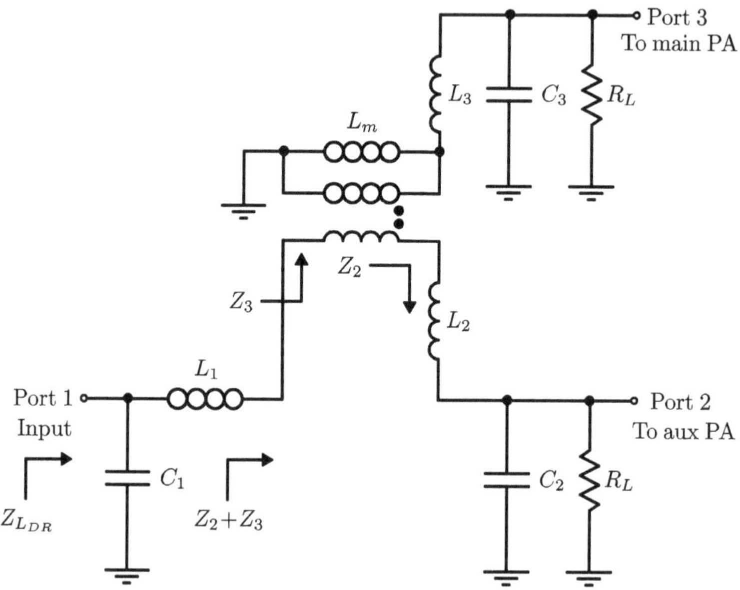

Fig. 4.14 Transformed-based 90° hybrid equivalent circuit model

From Fig. 4.13, the combiner and splitter networks as well as the quarter-wave transmission lines present problems for silicon substrates. Moreover, Doherty amplifiers are fundamentally difficult to implement at mm-wave frequencies because of the matching complexity and load pulling between the two PAs.

Yu et al. suggest a transformer-based quadrature hybrid that is implemented in a 65 nm bulk CMOS process. The hybrid replaces the input splitter and quarter-wave transmission line, reducing footprint size and power loss. Figure 4.14 shows the equivalent lumped-element circuit [36].

The impedance of the port 3 network is in series between the port 1 and port 2 networks, and the impedance looking into port 2 is

$$Z_2 = R_L || \frac{1}{j\omega C_2} + j\omega L_2 \tag{4.6}$$

Similarly, the impedance looking into port 3 is

$$Z_3 = \left(R_L || \frac{1}{j\omega C_3} + j\omega L_3 \right) || j\omega L_m \tag{4.7}$$

The final design occupies $0.11 \times 0.06\ \text{mm}^2$ and the hybrid provides excellent flexibility to fine-tune the impedance.

4.3 Filters

High-performance passive circuits are vital for the success of mm-wave designs, and filters are no different. Implementing filters alongside active components improves system integrity and reliability, primarily through avoiding the requirement for sophisticated packaging techniques. On-chip filters are perhaps the most widely used passive circuit block.

III-IV technologies offer unparalleled (i.e. lowest) insertion loss performance but suffer from high fabrication costs and poor integration capability with the digital baseband. This limits the potential for complete on-chip solutions in III-IV technologies. On the other hand, silicon provides excellent integration capabilities at the cost of higher passive losses due to the substrate properties. Lower-performance passive devices—especially in terms of insertion loss and harmonic suppression—limit the overall performance of any wireless IC. For example, cascading multiple lower-order bandpass filters (BPFs) can potentially improve harmonic suppression but significantly deteriorates the insertion loss. Successfully balancing the tradeoff between chip area and performance is a challenging task.

4.3.1 State-of-the-Art Bandpass Filters

Table 4.1 shows a summary of state-of-the-art BPF designs.

Table 4.1 State-of-the-art mm-wave filters

Ref	f_c (GHz)	Technology	Insertion Loss (dB)	Stopband attenuation (dB)	3dB BW (%)	Area (mm^2)
[5]	65	0.18μm CMOS	2.7	15	51.3	0.15
[37]	60, 77	0.18μm CMOS	9.3	40	10	0.11
[38]	35	0.18μm CMOS	4.5	35	37.8	0.124
[39]	31	0.13μm SiGe BiCMOS	2.4	20	22.6	0.024
[7]	31	0.13μm SiGe BiCMOS	3.9	45	51	0.075
[7]	35	0.13μm SiGe BiCMOS	3.1	35	50	0.075
[40]	33	0.13μm SiGe BiCMOS	2.6	20	42.4	0.031
[41]	29	0.13μm SiGe BiCMOS	3.5	47	26.7	0.028

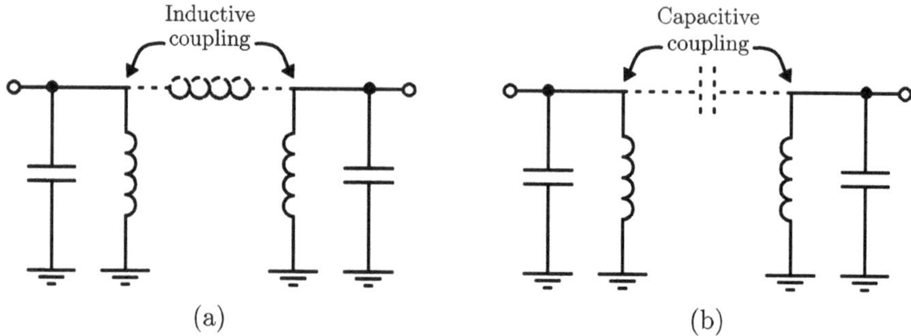

Fig. 4.15 Inductive and capacitive coupling models of a conventional bandpass filter

4.3.2 Silicon Integrated Passive Device Filters

Figure 4.15 shows two approaches to bandpass filter design. Conventional bandpass filters are generally implemented with inductive coupling [42]. Such filters are reasonably easy to design but are subject to sizing issues because of the coupling inductor.

The alternative capacitively coupled design is advantageous since the coupling capacitor is significantly smaller, even though it suffers from greater parasitics. To address the problems with the LC resonator, a good option is to minimise the distance between the inductive traces. In doing so, the resonator coupling is implemented with stray capacitance and mutual inductance instead of the coupling between L and C. Figure 4.16 shows the improved filter structure (a) and its equivalent circuit (b).

The integrated passive device (IPD) filter reported by Kim et al. is implemented in a unique IPD process on a lossy silicon substrate [42]. First, a 0.2 μm Au layer is deposited on silicon, forming the ground layer. The metal ground layer isolates signal traces from the substrate and it also means that classical microstrip filter

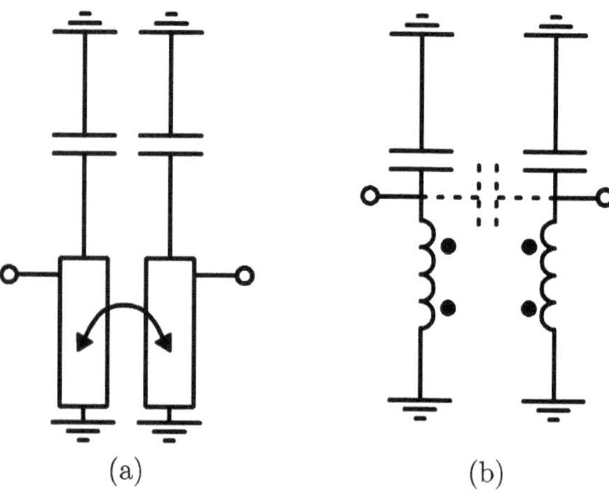

Fig. 4.16 **a** Improved bandpass filter design with line inductors and **b** its equivalent circuit

design procedures can be followed. Next, SiNx is deposited onto the metal ground layer via plasma-enhanced chemical vapour deposition, forming the isolating layer that acts as the insulator for the MIM capacitor. The SiNx layer is also the target of mask patterning, where plasma etching is used to remove areas needed for ground connections. Metal seed is then similarly deposited to form the top layer of the MIM capacitor.

Increasing the thickness of the isolating layer means that the optimal linewidth used for inductor synthesis increases, subsequently reducing filter insertion loss.

4.3.3 *Broadside-Coupled Resonator BPF*

Zhu et al. propose a method for synthesising higher-order bandpass filters on-chip by using broadside-coupled resonators in 130 nm SiGe BiCMOS [5, 41, 43].

4.3.3.1 Resonator Theory

The proposed approach is based on cells that can be expanded to realise higher order filters. Cells consist of broadside-coupled resonators with MIM capacitors to form a T-network. The layout of a broadside-coupled resonator is shown in Fig. 4.17.

The two meandered lines in Fig. 4.17 are, as the name implies, coupled in their broadside direction. Both have identical dimensions, and ground pads are connected through a shielding ground. Figure 4.18 shows the equivalent circuit.

First and second layer metal strips are represented by L_1, coupling between metal traces is C_1 and the coupling to ground is C_2. Capacitances are determined by the metal layer thickness as well as the dimensions and separation of the traces. Both ground capacitances (C_2) are assumed equal since their magnitude is much smaller than the coupling capacitance values. The symmetry in the circuit shown in Fig. 4.18

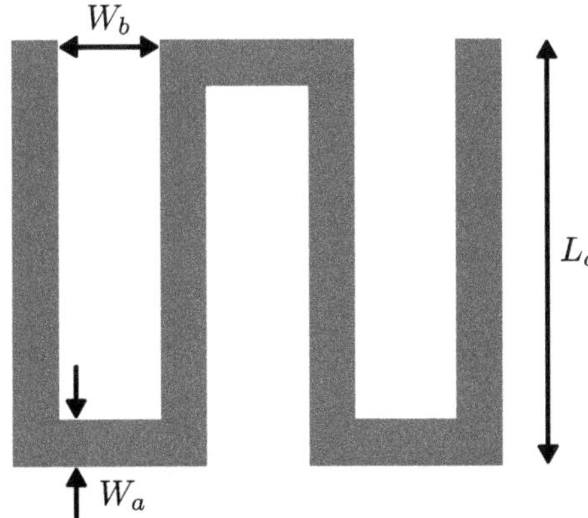

Fig. 4.17 Broadside-coupled resonator structure

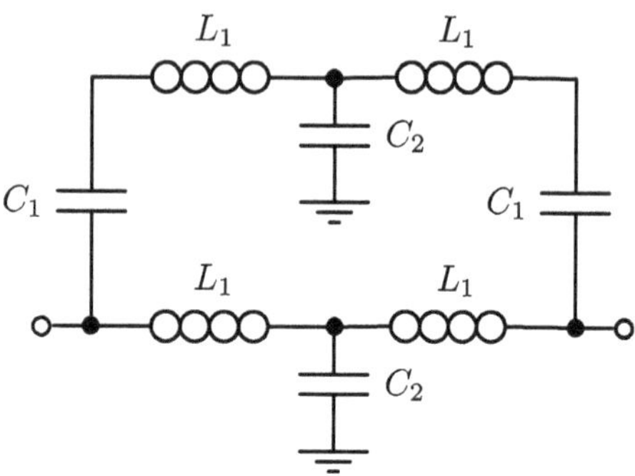

Fig. 4.18 Broadside-coupled resonator equivalent circuit

means that even/odd-mode analysis can be applied and consequently used to derive the S-parameters [24].

The even and odd-mode admittances are

$$Y_{even} = -\frac{j}{\omega L_1} + \frac{j\omega C_1}{1 - \omega^2 L_1 C_1} \tag{4.8}$$

and

$$Y_{odd} = \frac{j\omega C_2}{1 - \omega^2 L_1 C_2/2} + \frac{j\omega C_1 C_2/2}{C_1 + C_2/2 - \omega^2 L_1 C_1 C_2/2} \tag{4.9}$$

The S-parameters of the two-port network are then

$$S_{21} = \frac{Y_0(Y_{odd} - Y_{even})}{(Y_0 + Y_{odd})(Y_0 + Y_{even})} \tag{4.10}$$

and

$$S_{11} = \frac{Y_0^2 - Y_{even}Y_{odd}}{(Y_0 + Y_{odd})(Y_0 + Y_{even})} \tag{4.11}$$

Directly connecting the two resonator ports means that the network operates as a simple notch filter. Solving for $S_{21} = 0$ yields the location of the notch. From (4.10), this simplifies to $Y_{even} - Y_{odd} = 0$. The zero may be found from (4.8) and (4.9), which results in

$$\omega_{tz}^2 = \frac{C_1 + C_2/2 - \sqrt{2\left(2C_1^2 - C_2^2\right)}}{2L_1 C_1 C_2} \tag{4.12}$$

Since C_2 is small compared to C_1, it has a minor effect on the location of the zero. The resonator bandwidth and loss are determined by its external quality factor, which is

$$Q_{ex} = \frac{2Y_0 g_0 g_1}{b_r} \tag{4.13}$$

where g_0 and g_1 are the element values of a low-pass prototype filter and b_r is approximated as

$$b_r = \frac{\omega_0 \cdot \partial Im\{Y_{even}(\omega_0) + Y_{odd}(\omega_0)\}}{4 \cdot \partial\omega} \tag{4.14}$$

4.3.3.2 Filter Design

The broadside-coupled resonator can be expanded into a bandpass filter with relative ease, but it requires effective control over the coupling mechanism. Figure 4.19 shows one such implementation. In Fig. 4.19, input and output admittance inverters are implemented with series capacitors (C_p) [7]. The additional shunt capacitors ($-C_{pa}$) are added to tune the value of the admittance inverter, resulting in an L-shaped network.

Additionally, $-C_{pa}$ is necessary to provide the appropriate bandpass quality factor. Another shunt capacitor (C_p) is then added to compensate for the effect of $-C_{pa}$. The filter reported by Zhu et al. is implemented in a 0.13 μm BiCMOS technology, particularly on the two top metal layers. The high operating frequency range (23.9 to 44.3 GHz) is high enough to allow some capacitances in the filter to

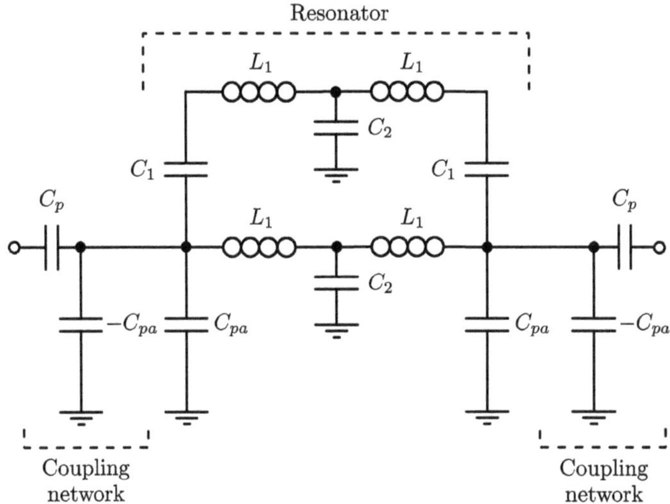

Fig. 4.19 LC-equivalent circuit of the resonator alongside its input and output coupling networks

be realised with parasitic components, which reduces the footprint of the cascaded design. The total chip area consumed by a third-order filter implemented with this method is 248 × 294 μm (excluding ground and signal pads used for probing).

References

1. Zheng L, Cui S, He K (2020) Overview of on-chip millimetre-wave passive circuit design in silicon-based technologies. In: IEEE MTT-S International Microwave Workshop Series on Advanced Materials and Processes for RF and THz Applications (pp 11–13). https://doi.org/10.1109/IMWS-AMP49156.2020.9199726
2. Jia H et al (2015) Research on CMOS mm-wave circuits and systems for wireless communications. China Commun 12(5):1–13. https://doi.org/10.1109/CC.2015.7112031
3. Camarchia V, Quaglia R, Piacibello A, Nguyen DP, Wang H, Pham AV (2020) A review of technologies and design techniques of millimeter-wave power amplifiers. IEEE Trans Microw Theory Tech 68(7):2957–2983. https://doi.org/10.1109/TMTT.2020.2989792
4. Chirala MK, Nguyen C (2006) Multilayer design techniques for extremely sminiaturised CMOS microwave and millimeter-wave distributed passive circuits. IEEE Trans Microw Theory Tech 54(12):4218–4224. https://doi.org/10.1109/TMTT.2006.885567
5. Sun S, Shi J, Zhu L, Rustagi SC, Mouthaan K (2007) Millimeter-wave bandpass filters by standard 0.18-μm CMOS technology. IEEE Electron Device Lett 28(3):220–222. https://doi.org/10.1109/LED.2007.891305
6. El-Hameed ASA, Barakat A, Abdel-Rahman AB, Allam A, Pokharel RK (2017) Ultracompact 60-GHz CMOS BPF employing broadside-coupled open-loop resonators. IEEE Microwave Wirel Compon Lett 27(9):818–820. https://doi.org/10.1109/LMWC.2017.2734771
7. Zhu H, Zhu X, Yang Y, Xue Q (2018) Design of wideband third-order bandpass filters using broadside-coupled resonators in 0.13-μm (Bi)-CMOS technology. IEEE Trans Microw Theory Tech 66(12):5593–5604. https://doi.org/10.1109/TMTT.2018.2873342
8. Razavi B (2006) A 60-GHz CMOS receiver front-end. IEEE J Solid-State Circuits 41(1):17–22. https://doi.org/10.1109/JSSC.2005.858626
9. Laskar J et al (2001) Development of integrated 3D radio front-end system-on-package (SOP). Gallium Arsenide Integrated Circuit (GaAs IC) Symposium, 2001. 23rd Annual Technical Digest (pp 215–218). https://doi.org/10.1109/GAAS.2001.964381
10. Chevalier P et al (2018) SiGe BiCMOS current status and future trends in Europe. In: IEEE BiCMOS and Compound Semiconductor Integrated Circuits and Technology Symposium (pp 64–71). https://doi.org/10.1109/BCICTS.2018.8550963
11. Marsh S (2006) Practical MMIC design. Artech House Inc., Boston, Massachusetts
12. Wagner E, Rebeiz GM (2019) Single and power-combined linear E-band power amplifiers in 0.12-μm SiGe with 19-dBm average power 1-GBaud 64-QAM modulated waveforms. IEEE Trans Microw Theory Tech 67(4):1531–1543. https://doi.org/10.1109/TMTT.2019.2893177
13. Hashemi H, Raman S (eds) (2016) mm-Wave silicon power amplifiers and transmitters. Cambridge University Press, Cambridge, United Kingdom
14. Datta K, Hashemi H (2017) Watt-level mm-wave power amplification with dynamic load modulation in a SiGe HBT digital power amplifier. IEEE J Solid-State Circuits 52(2):371–388. https://doi.org/10.1109/JSSC.2016.2622710
15. Datta K, Roderick J, Hashemi H (2013) Analysis, design and implementation of mm-Wave SiGe stacked Class-E power amplifiers. In: IEEE Radio Frequency Integrated Circuits (RFIC) Symposium (pp 275–278). https://doi.org/10.1109/RFIC.2013.6569581
16. Daneshgar S et al (2020) High-power generation for mm-Wave 5G power amplifiers in deep submicrometer planar and FinFET bulk CMOS. IEEE Trans Microw Theory Tech 68(6):2041–2056. https://doi.org/10.1109/TMTT.2020.2990638

17. Pekarik JJ et al (2014) A 90nm SiGe BiCMOS technology for mm-wave and high-performance analog applications. In: IEEE Bipolar/BiCMOS Circuits and Technology Meeting (pp 92–95). https://doi.org/10.1109/BCTM.2014.6981293
18. Doan CH, Emami S, Niknejad AM, Brodersen RW (2005) Millimeter-Wave CMOS design. IEEE J Solid-State Circuits 40(1):144–154. https://doi.org/10.1109/JSSC.2004.837251
19. Hajimiri A, Hashemi H, Natarajan A, Guan X, Komijani A (2005) Integrated phased array systems in silicon. Proc IEEE 93(9):1637–1654. https://doi.org/10.1109/JPROC.2005.852231
20. Lee CH et al (2002) A compact LTCC-based Ku-band transmitter module. IEEE Trans Adv Packag 25(3):374–384. https://doi.org/10.1109/TADVP.2002.805315
21. Datta K, Hashemi H (2014) Performance limits, design and implementation of mm-Wave SiGe HBT Class-E and stacked Class-E power amplifiers. IEEE J Solid-State Circuits 49(10):2150–2171. https://doi.org/10.1109/JSSC.2014.2353800
22. Mediano A, Molina-Gaudó P, Bernal C (2007) Design of class E amplifier with nonlinear and linear shunt capacitances for any duty cycle. IEEE Trans Microw Theory Tech 55(3):484–491. https://doi.org/10.1109/TMTT.2006.890512
23. Datta K, Hashemi H (2017) High-breakdown, high–fmax multiport stacked-transistor topologies for the W-Band power amplifier. IEEE J Solid-State Circuits 52(5):1–15. https://doi.org/10.1109/JSSC.2016.2641464
24. Pozar DM (2012) Microwave engineering, 4th edn. John Wiley & Sons Inc., Hoboken, New Jersey
25. Marsh SP (1997) MMIC power splitting and combining techniques. In: IEE Tutorial Colloquium on Design of RFIC's and MMIC's (pp 6/1–6–7)
26. Wu K, Lai K, Hu R, Jou CF, Niu D, Shiao Y (2014) 77–110 GHz 65-nm CMOS power amplifier design. IEEE Trans Terahertz Sci Technol 4(3):391–399
27. du Preez J, Sinha S (2017) Millimeter-wave power amplifiers. Springer International Publishing
28. Datta K, Roderick J, Hashemi H (2012) A 22.4 dBm two-way Wilkinson power-combined Q-band SiGe class-E power amplifier with 23% peak PAE. In: IEEE Compound Semiconductor Integrated Circuit Symposium (pp 5–8). https://doi.org/10.1109/CSICS.2012.6340076
29. Franc AL, Pistono E, Gloria D, Ferrari P (2012) High-performance shielded coplanar waveguides for the design of CMOS 60-GHz bandpass filters. IEEE Trans Electron Devices 59(5):1219–1226. https://doi.org/10.1109/TED.2012.2186301
30. Parveg D, Vahdati A, Varonen M, Karaca D, Karkkainen M, Halonen KAI (2015) Modeling and applications of millimeter-wave slow-wave coplanar coupled lines in CMOS. In: 10th European Microwave Integrated Circuits Conference Modeling (pp 207–210). https://doi.org/10.1109/EuMIC.2015.7345105
31. Parveg D, Varonen M, Karaca D, Halonen K (2019) Wideband mm-Wave CMOS slow wave coupler. IEEE Microwave Wirel Compon Lett 29(3):210–212. https://doi.org/10.1109/LMWC.2019.2892845
32. Parveg D, Varonen M, Karaca D, Vahdati A, Kantanen M, Halonen KAI (2018) Design of a D-band CMOS amplifier utilizing coupled slow-wave coplanar waveguides. IEEE Trans Microw Theory Tech 66(3):1359–1373. https://doi.org/10.1109/TMTT.2017.2777976
33. Bowers SM, Sengupta K, Dasgupta K, Parker BD, Hajimiri A (2013) Integrated self-healing for mm-wave power amplifiers. IEEE Trans Microw Theory Tech 61(3):1301–1315. https://doi.org/10.1109/TMTT.2013.2243750
34. May JW, Rebeiz GM (2010) Design and characterisation of W-band SiGe RFICs for passive millimeter-eave imaging. IEEE Trans Microw Theory Tech 58(5 PART 2):1420–1430. https://doi.org/10.1109/TMTT.2010.2042857
35. Cripps SC (2006) RF power amplifiers for wireless communications, 2nd edn. Artech House Inc., Dedham, Massachussets
36. Yu C, Feng J, Zhao D (2021) A 28-GHz Doherty power amplifier with a compact transformer-based quadrature hybrid in 65-nm CMOS. IEEE Trans Circuits Syst II Express Briefs 68(8):2790–2794. https://doi.org/10.1109/TCSII.2021.3068473
37. Nan L, Mouthaan K, Xiong YZ, Shi J, Rustagi SC, Ooi BL (2008) Design of 60- and 77-GHz narrow-bandpass filters in CMOS technology. IEEE Trans Circuits Syst II Express Briefs 55(8):738–742. https://doi.org/10.1109/TCSII.2008.922427

38. Yeh LK, Chen CY, Chuang HR (2010) A millimeter-wave CPW CMOS on-chip bandpass filter using conductor-backed resonators. IEEE Electron Device Lett 31(5):399–401. https://doi.org/10.1109/LED.2010.2043333
39. Chakraborty S et al (2016) A broadside-coupled meander-line resonator in 0.13-μm SiGe technology for millimeter-wave application. IEEE Electron Device Lett 37(3):329–332. https://doi.org/10.1109/LED.2016.2520960
40. Zhong Y, Yang Y, Zhu X, Dutkiewicz E, Shum KM, Xue Q (2017) An on-chip bandpass filter using a broadside-coupled meander line resonator with a defected-ground structure. IEEE Electron Device Lett 38(5):626–629. https://doi.org/10.1109/LED.2017.2690283
41. Bautista MG, Zhu H, Zhu X, Yang Y, Sun Y, Dutkiewicz E (2019) Compact millimeter-wave bandpass filters using quasi-lumped elements in 0.13-um (Bi)-CMOS technology for 5G wireless systems. IEEE Trans Microw Theory Tech 67(7):3064–3073. https://doi.org/10.1109/TMTT.2019.2895581
42. Kim DM, Min BW, Yook JM (2019) Compact mm-wave bandpass filters using silicon integrated passive device technology. IEEE Microwave Wirel Compon Lett 29(10):638–640. https://doi.org/10.1109/LMWC.2019.2936688
43. Sun F, Zhu H, Zhu X, Yang Y, Sun Y, Zhang X (2019) Design of millimeter-wave bandpass filters with broad bandwidth in Si-based technology. IEEE Trans Electron Devices 66(3):1174–1181. https://doi.org/10.1109/TED.2019.2895161

Chapter 5
Solid-State Millimeter-Wave Silicon Amplifiers

Amplifiers are crucial components in any wireless transceiver MMIC and are used extensively. Power amplifiers (PAs) are the last components in the transmitter chain, while low-noise amplifiers (LNAs) are usually the first components in the receiver chain. Silicon amplifiers offer some distinct advantages over their III-IV counterparts. For one, seamless integration with digital circuits on the same substrate is a major bonus and is often regarded as the most significant advantage of CMOS and BiCMOS technologies. BiCMOS, specifically, takes the integration capability one step further by adding high-performance analogue building blocks and passives into the mix. As such, these technologies are well worth consideration for mm-wave PA and LNA implementations.

5.1 Amplifier Specifications

This chapter starts with a high-level discussion on typical amplifier performance characteristics. Most of the metrics discussed here are applicable to LNA and PA design, albeit perhaps with differing objectives.

5.1.1 Gain and Stability

MMIC amplifier design involves careful consideration of the gain required from the whole chip relative to the gain attainable from each amplifier stage. The maximum achievable gain is inhibited purely by the necessity for stability. Looking at the MMIC as a whole, this can be achieved by keeping the chip gain below the gain from any feedback paths from the output of the chip to its input. While there are many possible feedback paths, the most common one is reverse isolation (or S_{12} for a single-port

J. du Preez and S. Sinha, *State-of-the-Art of Millimeter-Wave Silicon Technology*,
Lecture Notes in Electrical Engineering 945,
https://doi.org/10.1007/978-3-031-14655-8_5

device). Additionally, paths through cables, power supplies and DC bias traces can be significant when RF decoupling is insufficient. MMICs mounted within metal enclosures are also susceptible to cavity waveguide modes [1].

Common-ground feedback can arise when there is a poor ground connection to the enclosure base plate, and it can also cause parallel-plate modes to propagate. RF probes may also cause interference through radiation. These factors make it extremely difficult to characterise feedback accurately. Moreover, feedback is inversely proportional to frequency, further complicating the task for mm-wave IC design.

Devices operating close to their f_T limits will experience some sort of inherent gain compression, which adds a degree of difficulty in mm-wave amplifier design.

5.1.1.1 Gain

Figure 5.1 shows a diagram of a generic two-port network, often used as the basis for amplifier specifications.

The reflection coefficients that are shown in Fig. 5.1 can be represented in terms of the network's S-parameters or its source and load impedances [2]. First, the input reflection coefficient is

$$\Gamma_{in} = \frac{Z_{in} - Z_0}{Z_{in} + Z_0} \tag{5.1}$$

And the output reflection coefficient is

$$\Gamma_{out} = S_{22} + \frac{S_{12}S_{21}\Gamma_s}{1 - S_{11}\Gamma_s} \tag{5.2}$$

Subsequently, complete expressions for the input power and the power delivered to the load can be obtained. The input power, in terms of the source voltage V_s is

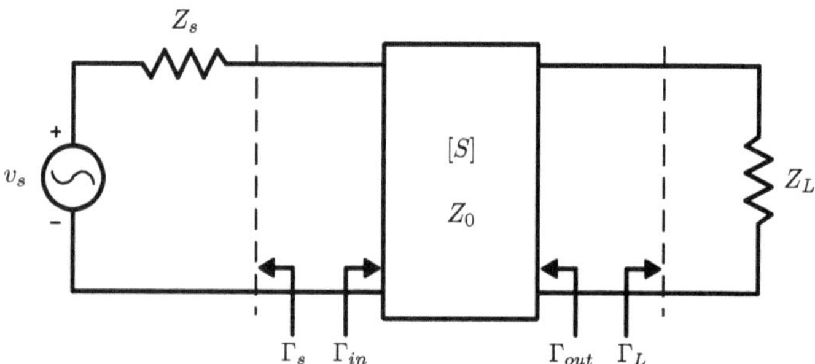

Fig. 5.1 Two-port network

$$P_{in} = \frac{|V_S|^2}{2Z_0}\left(\frac{|1-\Gamma_s|^2}{|1-\Gamma_s\Gamma_{in}|^2}\right)(1-\Gamma_{in}^2) \tag{5.3}$$

and the load power is

$$P_L = \frac{|V_S|^2}{8Z_0}\left(\frac{|S_{21}|^2\{1-\Gamma_L^2\}|1-|\Gamma_S|^2|}{|1-S_{22}\Gamma_L|^2|\Gamma_S\Gamma_L|^2}\right) \tag{5.4}$$

Finally, the amplifier gain is

$$G = \frac{P_L}{P_{in}} = \frac{|S_{21}|^2(1-\Gamma_L^2)}{|1-S_{22}\Gamma_L|^2(1-\Gamma_{in}^2)} \tag{5.5}$$

Other gain values used are the available power gain and the transducer power gain, but the only power gain is used for our discussions.

5.1.1.2 Stability

PA stability over its intended operating bandwidth is critical. The two-port network in Fig. 5.1 will oscillate when $|\Gamma_{in}| > 1$ or when $|\Gamma_{out}| > 1$, which makes intuitive sense since the reflected power exceeds the incident power in either case. The easiest way for these conditions to occur is when the impedances contain negative real parts. Moreover, Γ_{in} and Γ_{out} are controlled through the input and output matching networks respectively. The stability conditions are written as

$$|\Gamma_{in}| = \left|\frac{S_{11}-\Gamma_L\Delta}{1-S_{22}\Gamma_L}\right| < 1 \tag{5.6}$$

and

$$|\Gamma_{out}| = \left|\frac{S_{22}-\Gamma_s\Delta}{1-S_{11}\Gamma_s}\right| < 1 \tag{5.7}$$

where Δ is the determinant of the S-parameter matrix,

$$\Delta = S_{11}S_{22} - S_{12}S_{21} \tag{5.8}$$

In most cases, the input and output reflection coefficients will satisfy their stability conditions over a limited range. As such, instabilities are tolerable if they fall far enough outside the operating bandwidth. Careful simulation of the load and source impedances is necessary to ensure stable operation.

The so-called $K-\Delta$ test is often used to determine unconditional stability. Rollet's stability criterion needs $K > 1$, where

$$K = \frac{1 - |S_{11}|^2 - |S_{22}|^2 + |\Delta|^2}{2|S_{12}S_{21}|} \tag{5.9}$$

and Δ is given by (5.8). The B_1 factor is also sometimes used to test for unconditional stability, and it is

$$B_1 = 1 + |S_{11}|^2 - |S_{22}|^2 - |\Delta|^2 \tag{5.10}$$

5.1.2 Linearity

Linearity is a crucial amplifier parameter that has a profound effect on performance. It is also dependent on the application and modulation used. For example, 60 GHz WLAN uses higher-order QAM schemes where the modulated signals have high peak-to-average-power ratios (PAPRs). As such, high linearity is needed to maintain the QAM constellation's integrity and stay within a specified error vector magnitude (EVM).

5.1.2.1 Gain Compression

Amplifier gain compression is specified as either the input or output 1 dB compression point. This occurs when the output power drops 1 dB below its linear characteristic. The input and output compression points are related,

$$OP_{1\text{dB}} = IP_{1\text{dB}} + G - 1\text{dB} \tag{5.11}$$

Amplifier datasheets typically list $OP_{1\text{dB}}$, and the input limit can be determined with (5.11). For signals using some form of amplitude modulation, the input level must be adjusted so that the peak amplitude does not cause the amplifier to go into gain compression. Figure 5.2 defines the 1 dB compression point.

5.1.2.2 Intermodulation Distortion

The well-known Taylor series expansion defines the response of a nonlinear network to an input signal v_i,

$$v_o = a_0 + a_1 v_1 + a_2 v_i^2 + \ldots \tag{5.12}$$

The series consists of DC, linear and squared components, denoted by a_0, a_1 and a_2, respectively, and additional higher-order terms. Mixers and frequency multipliers

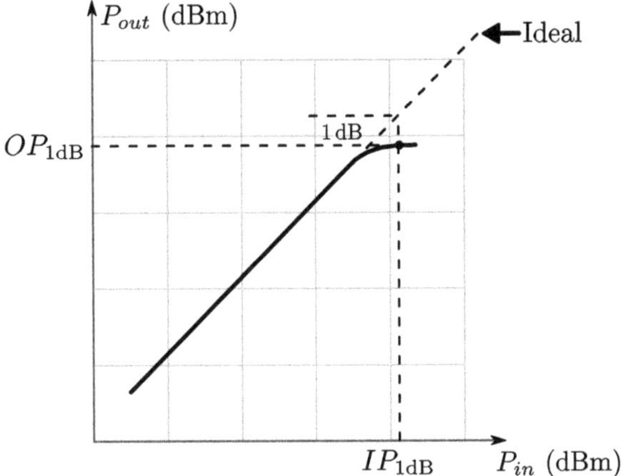

Fig. 5.2 Amplifier 1 dB compression point definition

are interested in higher-order terms, while amplifier design mainly involves isolating the linear component.

While single-tone inputs will result in harmonics that fall outside the band of interest, the situation is more difficult when two-tone inputs are present. In this case, harmonics and intermodulation products will appear as $m\omega_1 + n\omega_2$, where the two frequencies are ω_1 and ω_2 and $m, n = 0, \pm 1, \pm 2, \ldots$. Intermodulation products arise in the form of difference products ($2\omega_1 - \omega_2$ and $2\omega_2 - \omega_1$) that can distort the desired spectrum. Moreover, these closely spaced components can be challenging to eliminate via filtering, so accurate characterisation is essential to predict their effects on linearity.

Intermodulation distortion is typically specified as the third-order intercept point (IP_3). Consider the input–output power plot in Fig. 5.3.

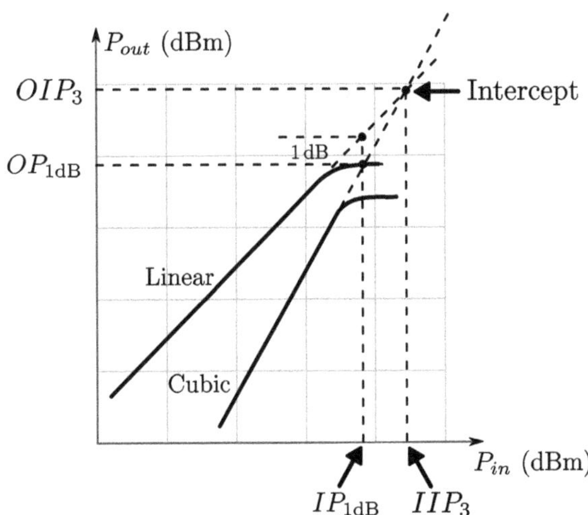

Fig. 5.3 Third-order intercept illustration

The ideal linear response of the third-order product, which is the cube of the input voltage, overlaps the linear response of the fundamental component at the intercept point. The input and output power levels where this occurs are IIP_3 and OIP_3, respectively. The cubic slope also means that weak input signals are largely unaffected by third-order products, but they become much more prominent as the input power is increased. As Fig. 5.3 shows, the amplifier experiences gain compression, and the first and third-order power curves will never coincide in practice.

5.1.2.3 Dynamic Range

Dynamic range is the signal range over which a component performs within specification. For LNAs, the lower limit of the dynamic range is due to the noise floor and the upper limit is located where intermodulation distortion is too severe. For PAs, the upper limit may be at the compression point. LNA datasheets will typically specify the spurious-free dynamic range (SFDR), while PA datasheets generally specify the linear dynamic range (LDR).

5.1.2.4 Error Vector Magnitude

Complex modulation schemes (that is, ones that use amplitude and frequency modulation) are used in communication systems to achieve greater spectral efficiency [3]. Moreover, PAs are nonlinear and naturally induce some non-ideal effects in the transmitted signal, which means that some method to determine modulation quality is required. Quantifying signal quality is most often done using the error vector magnitude (EVM) [4]. Multiple methods to calculate EVM are employed, and unfortunately, the results produced are not identical. As such, a consistent methodology must be maintained to generate a valid comparison between competing designs.

EVM is best described graphically alongside Fig. 5.4, showing a partial constellation diagram for a 16-QAM signal.

EVM measures the deviation from the ideal constellation point. It is arguably more helpful than bit error rate (BER), which only measures whether received bits are correct or not, seeing that it assigns a numeric value to signal quality. In Fig. 5.4, three vectors are shown, namely S_{max}, $S_{ideal,i}$ and $S_{meas,i}$. The first of these represent the maximum vector, which is $1 + j$ in this case. The second represents the ideal value for the ith constellation point, or $1 + j0.5$ in Fig. 5.4, and accompanying it is $S_{meas,i}$, the value for a measured point relating to $S_{ideal,i}$.

RMS and peak EVM are encountered most often in the literature [4]. First, RMS EVM is a ratio of RMS magnitudes, as given by

$$EVM_{RMS} = \frac{\sqrt{\frac{1}{N}\sum_i^N \left|S_{ideal,i} - S_{meas,i}\right|^2}}{\sqrt{\frac{1}{N}\sum_i^N \left|S_{ideal,i}\right|^2}} \tag{5.13}$$

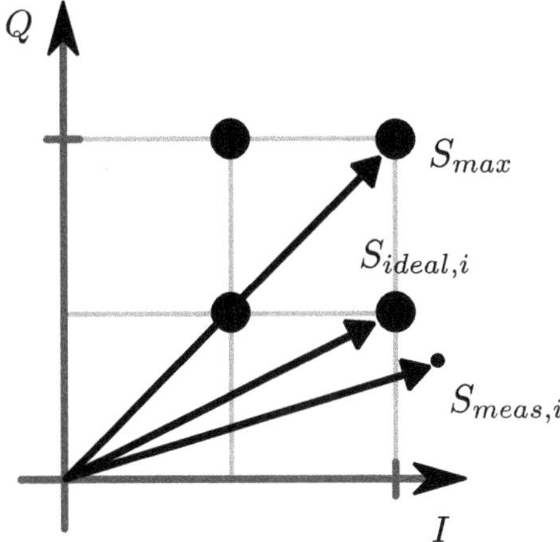

Fig. 5.4 Partial constellation diagram for a 16-QAM signal

The second approach normalises the RMS EVM by the peak magnitude of the constellation (S_{max}), and can be computed as

$$EVM_{max} = \frac{\sqrt{\frac{1}{N}\sum_i^N \left|S_{ideal,i} - S_{meas,i}\right|^2}}{|S_{max}|} \tag{5.14}$$

It is clear from (5.13) and (5.14) that the two approaches differ in their normalisation of the RMS error. Constellations where the magnitude is constant (as is the case with phase-modulated signals such as QPSK and BPSK), will result in the same EVM as computed with the above equations, while $EVM_{RMS} > EVM_{max}$ when amplitude modulation is also present (such as QAM and APSK).

5.1.3 Bandwidth

The bandwidth specification of an amplifier is determined from the target application. For example, E-band satellite communication operates in the $71-76$ GHz and $81-86$ GHz bands for uplink and downlink, respectively.

Large bandwidths are a primary attraction for operation in mm-wave bands. A 1% bandwidth at 6 GHz is merely 60 MHz, but the same percent-bandwidth is 600 MHz at 60 GHz.

5.1.4 Efficiency

Efficiency is loosely defined as the amplifier's ability to convert DC power into RF power. LNAs operating with very low signal levels at their input ports generally do not optimise efficiency. On the other hand, efficiency is crucial in PA design. The PA consumes the most power in the front-end, and poor efficiency degrades battery life and can lead to longevity issues for the transceiver. Unfortunately, efficiency and linearity, while more or less important depending on the application, are essentially contradicting requirements. Continuous-mode amplifiers exhibit good linearity but poor efficiency. The situation is reversed for switch-mode amplifiers.

A useful method to express efficiency is by including the effect of the amplifier gain, leading to its power-added efficiency (PAE)

$$PAE = \frac{P_{RF}}{P_{DC}}\left(1 - \frac{1}{G}\right) \tag{5.15}$$

5.1.5 Noise

Noise is typically represented by an average value of the mean-square variation around an average voltage or current [5]. This can be illustrated mathematically as

$$\overline{i^2} = \overline{\left(I - I_{avg}\right)^2} \tag{5.16}$$

where:

- $\overline{i^2}$ is the mean-square value of the current fluctuation
- I is the current fluctuation
- I_{avg} is the average current

Voltage noise sources are described similarly. Additionally, noise voltages and currents are also characterised over a specified bandwidth.

5.1.5.1 Thermal Noise

The thermal motion of charge carriers within any type of conductor will induce a random voltage at any temperature other than absolute zero. This is known as Johnson–Nyquist or thermal noise. This voltage is given by

$$V_t = \sqrt{4kBTR} \tag{5.17}$$

where:

- V_t is the thermal noise
- $k = 1.38 \times 10^{-23}$J/K is Boltzmann's constant
- B is the measurement bandwidth
- R is the resistance in Ohms

If a current flows in the resistor, this will contribute to the total noise in the form of shot noise.

5.1.5.2 Shot Noise

Shot noise consists of random current fluctuations caused by the random nature of discrete electrons. Increasing the average current magnitude increases the shot noise proportionally. The current flowing across a diode is comprised of holes from the p-region and electrons from the n-regions, both of which have sufficient energy to cross the barrier junction potential. Occasionally, randomness can cause carriers to cross the junction, inducing shot noise. Shot noise is expressed as

$$\overline{i^2} = 2qI_D B \tag{5.18}$$

where:

- I_D is the (forward) diode current
- B is the bandwidth
- $q = 1.6 \times 10^{-19}$°C is the electron charge

The expression for shot noise in (5.18) is only valid until the operating frequency reaches $1/\tau$, where τ is the depletion region carrier transfer time.

5.1.5.3 Flicker Noise

Flicker noise is present primarily in active devices. The traps created by crystal defects and contamination during fabrication are the leading cause of flicker noise. It has a $1/f$ dependence, and is sometimes referred to as $1/f$ noise. Its current is

$$\overline{i^2} = K_1 \frac{I^a}{f^b} B \tag{5.19}$$

where K_1, a and b are technology parameters. When all three are unity, the spectral density of the flicker noise current (normalised by bandwidth B) decreases linearly.

5.1.5.4 Transistor Noise Modeling

Transistor noise models are typically derived from their physical structure and subsequently inserted into their respective small-signal models. This sub-section will briefly highlight noise modelling approaches for bipolar and FET devices.

MOSFET Noise

MOSFETs are affected by two noise sources. First, the channel resistance changes with the device's transconductance g_m. As a result, since the channel current I_D also changes with g_m, thermal noise current will also appear. Provided that the frequency is sufficiently high for flicker noise to have no effect, the thermal noise current is

$$\overline{i_d^2} = 4kTB(K_2 g_m) \tag{5.20}$$

where $K_2 = 2/3$ for long-channel devices but is lower for the short-channel ones used in mm-wave circuits. The thermal noise current is, therefore, dependent on the process technology. The remainder of the parameters have already been defined.

The second noise source is generated from gate currents. Leakage current is the only component of the gate current at DC, so that the shot noise is negligibly small. However, the situation changes once thermal noise through the gate becomes dominant at higher frequencies. The capacitive reactance of C_{GS} ($X_{GS} = 1/\omega C_{GS}$) determines the current,

$$\overline{i_g^2} = K_3 kTB \cdot \frac{\omega^2 C_{GS}^2}{5 g_{d0}} \tag{5.21}$$

where:

- K_3 is another technology parameter that is equal to unity for large-channel FETs and that increases dramatically for short-channel devices.
- g_{d0} is the drain-source conductance when $V_{DS} = 0$

The two noise current sources are related since the FET gate is closely associated with the channel. Taking these into account, the small-signal model can be updated, as shown in Fig. 5.5.

HBT Noise

Operating bipolar transistors at sufficiently high frequencies also means that flicker noise can be neglected. The two pn junctions in bipolar devices means that there are two origins of shot noise. For a transistor operating in its forward-active mode, the base and collector currents will induce shot noise components,

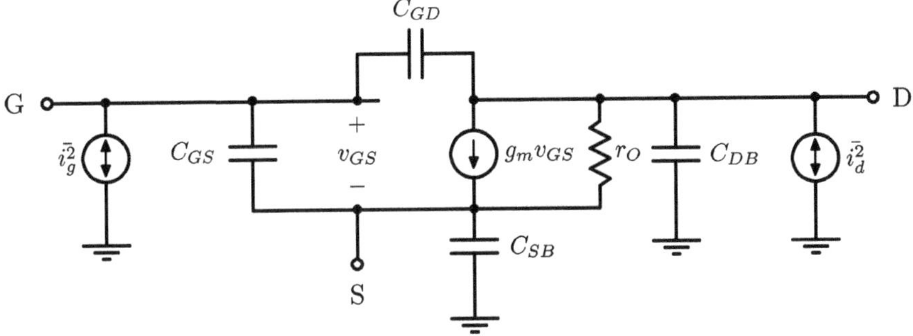

Fig. 5.5 Small-signal integrated MOSFET model with added noise sources

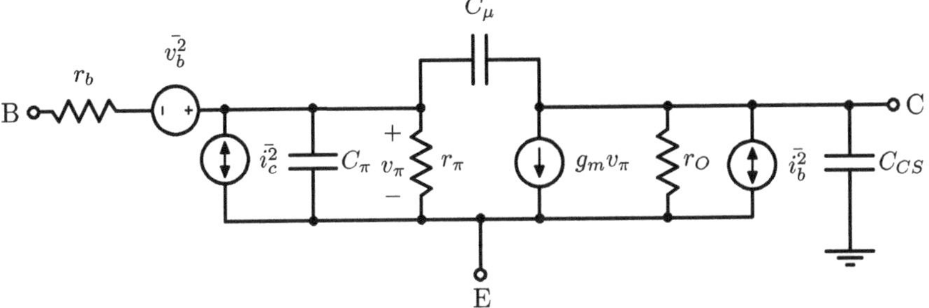

Fig. 5.6 Bipolar small-signal model with noise sources

$$\overline{i_b^2} = 2qI_B B \tag{5.22}$$

and

$$\overline{i_c^2} = 2qI_C B \tag{5.23}$$

Since r_π and r_o are not physical, they have no contributing noise currents. The base resistance r_b is also a source of thermal noise, and it is written as (Fig. 5.6)

$$\overline{v_b^2} = 4kTr_b B \tag{5.24}$$

5.1.5.5 Definition of Amplifier Noise

Transistor noise is characterised by referring the output thermal and shot noise back to an equivalent noise figure at its input. Consider the amplifier with gain G_A and noise figure N_A shown in Fig. 5.7.

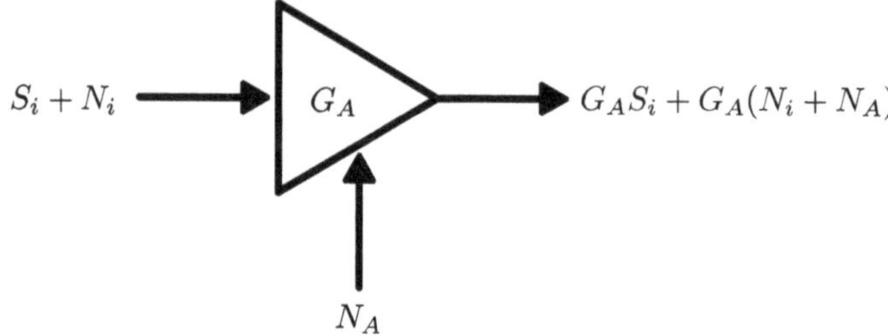

Fig. 5.7 Amplifier noise illustration

As Fig. 5.7 illustrates, the amplifier gain equally affects the input signal (S_i), the input noise (N_i) and the inherent amplifier noise (N_A). As such, the noise factor can be determined with

$$\frac{\text{total noise}}{\text{source noise}} = \text{noise factor}\ \frac{S_o/N_o}{S_i/N_i} = \frac{N_i + N_A}{N_i} \tag{5.25}$$

Subsequently, the noise figure is

$$\text{Noise figure (dB)} = 10 \times \log(\text{noise factor}) \tag{5.26}$$

Impedance matching plays a role in achieving the optimal noise figure, and it involves matching for Γ_{opt}. In this case, the noise figure is denoted by F_{min} and the gain is referred to as the associated gain (G_{Ass}).

The noise figure of the first component in a cascade of building blocks has the most significant effect on the overall noise figure, hence the importance of minimising the LNA's noise figure [2].

5.2 Amplifier Classification

Amplifiers are categorised according to their class of operation. Another approach often employed is to separate amplifiers based on their bias conditions, which, in turn, determines the current conduction angle in some amplifiers. The challenge is that the amplifier's operating class can be assigned from several distinct characteristics and can occasionally become ambiguous. However, the conduction angle approach has its limitations. Consider, for example, Class C amplifiers, where the conduction angle varies alongside input drive. Class AB amplifiers exhibit the opposite effect. Neither of these effects occurs in purely Class A and B amplifiers.

A better strategy for grouping amplifiers is to consider their dynamic operating conditions alongside the matching network termination [6]. Following this strategy, amplifiers are split into continuous and switched-mode circuits. In the former class, the active device is a current source, so FETs will be modelled as voltage-controlled current sources. The latter class assumes the transistor is a switch, and most analytical techniques treat the device as an ideal switch. Non-ideal switching effects are well documented [7]. Table 5.1 summarises a few key differences between switching and continuous-mode amplifiers.

A summary of amplifier classes is shown in Fig. 5.8. While there are other classes of operation (e.g. Class DE and inverse Class F), the ones shown in Fig. 5.8 are perhaps the most common [6].

Table 5.1 Comparison of switch-mode and continuous-mode amplifiers

Characteristic	Switch-mode	Continuous-mode
AC output impedance	Low during on-cycle, high during off-cycle	High
Device driven into saturation?	Yes, in some cases	No
Device voltage during conduction cycle	Device voltage minimised to limit the effects of voltage-current overlap	Voltage rises above a minimum point, which depends on the output network's response to pulsed currents
Output network design criteria	Input voltage determined by altering between open and short-circuited connections	Input voltage determined by response to current pulse trains
Load network input impedance affects load network?	Yes	No
Factors that influence conduction current	Determined by the load network	Purely dependent on drive current

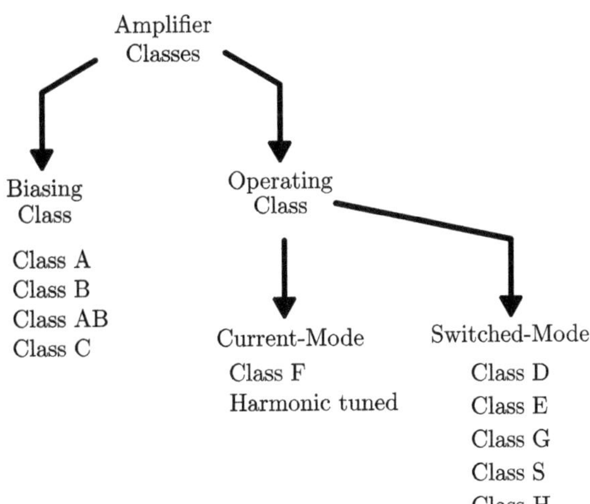

Fig. 5.8 Amplifier classification

5.3 Low-Noise Amplifiers

LNA design assumes that the active devices' RF levels are sufficiently low to be entirely linear. Therefore, standard S-parameter techniques can be used to design LNAs, which can be iterated rapidly with simulation tools. LNAs are used in receivers to amplify very low signal levels from the antenna (or the band select filter) while adding the least noise possible. The gain of the LNA reduces the contribution of subsequent components to the total noise figure.

For some time, high-performance LNAs operating in mm-wave bands were not realisable. Instead, receivers used mixers after the antenna to first downconvert incoming signals to a more usable frequency band. The apparent consequence of this is that the system noise figure is markedly worse when compared to using an LNA directly. Moreover, this mixer introduced conversion losses, which hampers the dynamic range and further degrades the noise figure for subsequent receiver components. The first mm-wave LNAs were fabricated in GaAs, and the demand for high-performance LNAs in GaAs, SiGe and CMOS is still growing today.

5.3.1 Millimeter-Wave Design Techniques

5.3.1.1 Performance Characterisation

The need for matching to Γ_{opt} at the LNA input was mentioned earlier in Sect. 5.1.5. Γ_{opt} is dependent on the bias level and the source (or emitter) termination. At lower frequencies, FET devices are optimally biased at around 10% of I_{dss}, which induces minimal gain degradation (the ideal value being at $0.5I_{dss}$) [8]. The low current may affect linearity and, consequently, dynamic range, but this is generally not an issue when the input signal levels are low.

However, LNA design goals differ at mm-wave frequencies, primarily due to limited transistor performance. Reducing the bias current may reduce the gain significantly, rendering the approach as mentioned above useless.

LNA design goals are characterised through a figure-of-merit (FoM) that relates the gain G to the noise factor F and power consumption P [9],

$$FoM_{LNA} = \frac{G}{(F-1)\cdot P} \tag{5.27}$$

The classical FoM in (5.27) is commonly expanded by adding frequency and IIP_3 terms to the numerator.

Operating devices at mm-wave frequencies close to f_T is anything but trivial since it is challenging to contend with the reduced gain and poor noise figure. Downscaling devices means that the breakdown voltages are reduced, reducing the supply voltages and limiting linearity. Belostotski and Klumperink report a new set of LNA FoMs as well as their practical limitations on CMOS technology [10]. In fact, at least 30 LNA FoMs have been used throughout the literature [11]. In most cases, a particular FoM is used to determine LNA performance compared to other designs, presumably on equal footing. The FoM characterises a particular design strategy relative to others, and it provides a means for comparing. However, there are problems with the amount of FoMs used. For one, a particular LNA may perform well against its competition based on a certain FoM, where the situation is reversed for another FoM. This makes objective comparison and genuine evaluation of technologies and design techniques difficult.

Another layer of complexity stems from the fact that different LNAs have different performance targets. Radio astronomy applications emphasise noise temperature, while power consumption is neglected. For IoT devices, power consumption is crucial. Should these LNAs be compared on the same FoM? It is easy to see why not. Moreover, different technologies are analysed with different FoMs, which begs the question, can we compare SiGe and CMOS LNAs with FoMs developed for CMOS LNAs? In some cases, perhaps yes, e.g. where technology parameters are not included in the calculation.

Given these problems, the comparisons drawn in this section will be based on basic LNA parameters. FoM analysis can be perused in greater detail in the reports by Belostoski and Jagtap [10–12].

Minimum Noise Factor and Noise Measure

The fundamental limit of noise factor for a MOSFET transistor is

$$F_{min} = 1 + \frac{2}{\omega_0 \omega_T} \sqrt{\frac{\delta \gamma}{5} \left(1 - |c|^5\right)} \tag{5.28}$$

where:

- δ and γ are coefficients that quantify the excess noise in the drain and gate currents
- c is the correlation coefficient between the abovementioned noise currents

The noise measure of the LNA (F_M) is

$$F_M = \frac{F - 1}{1 - \frac{1}{G}} \tag{5.29}$$

where:

- F is the LNA noise figure
- G is the LNA gain

The lowest possible noise figure in a cascaded system is achieved when the LNA with the lowest F_M (not the lowest F) is positioned first [10]. Furthermore,

- The minimum F_M is independent of device size
- F_M of any number of cascaded two-port stages is never less than the best-case F_M
- F of a whole network can never be less than that of the lowest-F stage
- Lossy or active embedding can only add noise and thus degrades the minimum possible F_M

Linearity Measure

Linearity measure is defined as

$$P_{Lin} = IP_{1\text{dB}}(G - 1) \tag{5.30}$$

Maximising $IP_{1\text{dB}}$ of a cascaded pair of amplifiers can be achieved by placing the amplifier with the higher P_{Lin} last in the chain. As a consequence, P_{Lin} characterises the gain-linearity tradeoff.

5.3.1.2 Design Methodology

Cascode topologies are typically employed for mm-wave LNA design since it offers better isolation as well as higher gain and bandwidth than individual common-source or common-gate stages. Figure 5.9 shows a cascode bipolar amplifier.

Adding an inductor between the common-emitter (or common-source) and common-base (or common-gate) stages cancels out the middle pole and acts as compensation for the lower f_T of the cascode pair [13, 14].

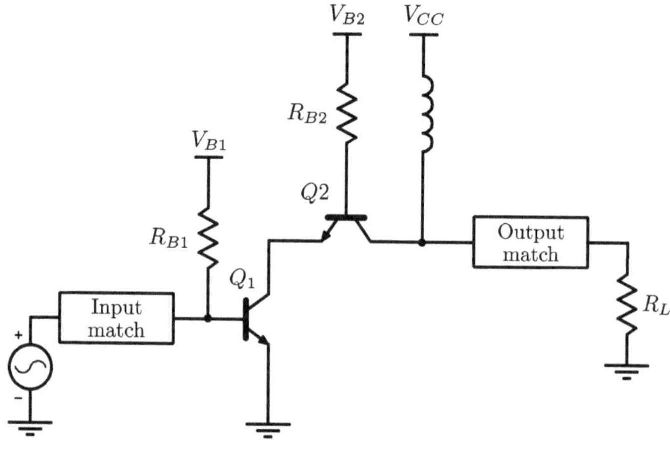

Fig. 5.9 Cascode bipolar amplifier

Yao et al. propose the following procedure for designing mm-wave CMOS cascode LNAs [9, 15].

1. Set the bias point so that the current density (J_{opt}) corresponds to the optimal noise figure, NF_{min}.
2. Size the transistor to produce the highest possible f_{max} and the minimal NF_{min}. The feature size, W_f, is somewhere between $1 - 2\,\mu$m for 90 nm CMOS.
3. Choose the best value of the interstage inductor (L_M) when the cascode pair is biased at J_{opt} by plotting f_T versus L_M.
4. Add additional transistor fingers to match the source impedance to the optimal noise impedance at the centre frequency.
5. Find the optimal value for the source inductor (L_G) after parasitic extraction.
6. Add an additional series gate inductor (L_G) to tune out the imaginary parts of the input impedance and the optimal source impedance.
7. Design the output matching network for maximum gain.

5.3.1.3 Millimeter-Wave Challenges

Designing mm-wave LNAs in SiGe or Si CMOS is challenging. Bulk CMOS processes targeted at lower power digital applications (such as portable wireless devices for the consumer market) are not optimised for RF design and pose several challenges [16]:

- The digital nature of the process means that RF device models are substandard and frequently have to be developed.
- The total height of the metal layer stackup is relatively thin compared to RF CMOS processes, leading to larger parasitic resistances and capacitances. It also means that planar inductors are very limited in terms of the achievable quality factor.
- Bulk technologies generally do not have dedicated MIM layers, which means that designers have to resort to VNCaps between metal layers or MOSFET capacitors. Neither approach can offer comparable high-Q capacitors with sufficiently high self-resonant frequencies, as do RF technologies.
- Deep submicron scaling means that breakdown voltages are very low, sometimes in the range of 1 V. Alongside this, layout rules are strict and complicated and large process variations are to be expected.

SiGe BiCMOS and RF CMOS technologies are advantageous here, but both come with cost implications. Additionally, in some cases, the digital IP blocks available in highly scaled CMOS technologies might not exist in comparable RF CMOS or BiCMOS processes.

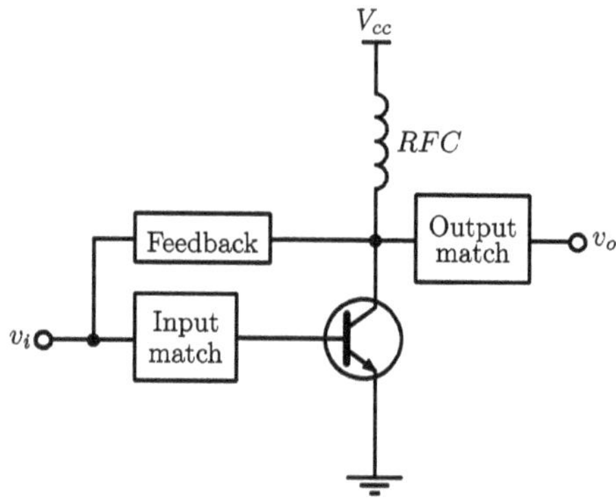

Fig. 5.10 Common-emitter amplifier

5.3.2 LNA Operating Characteristics

LNAs are constructed in common-emitter (or common-source) configurations, as shown in Fig. 5.10.

Other configurations (common-base and cascode) are also employed. The cascode configuration is attractive because it allows for independent input and output matching and can achieve higher gain [17]. The input match is designed to achieve optimal noise and power matching simultaneously. Transistor parasitics will cause a portion of the output to appear at the input.

LNA power gain is defined as

$$G = \frac{P_L}{P_{in}} \tag{5.31}$$

which is simply the ratio between load and input power. Conversely, reverse isolation is simply the opposite. Maximising G is desirable, while reverse isolation must be minimised. In LNA design, the gain is often sacrificed somewhat in order to improve noise figure performance.

5.3.3 State-of-the-Art Silicon LNAs

Figure 5.11 shows a comparison in noise figure performance between SiGe and CMOS LNAs operating above 20 GHz.

At higher frequencies (above 75 GHz), Fig. 5.11 shows that SiGe LNAs tend to provide better noise figure performance. Performance is similar below 75 GHz, with CMOS LNAs slightly ahead on average.

Song et al. report a 95 GHz LNA implemented in 90 nm SiGe BiCMOS [18]. This LNA achieves 34 dB gain with a minimum noise figure of 3.5 dB (at 80 GHz). The noise figure is below 4.5 dB over the entire W-band range. This LNA operates

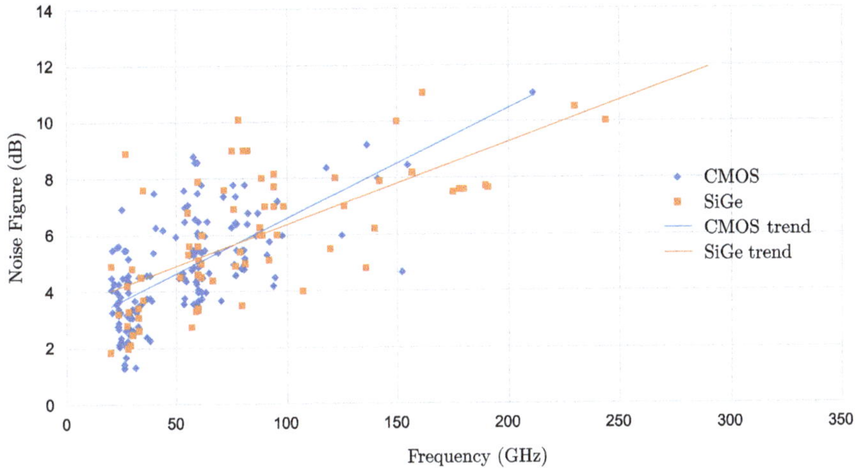

Fig. 5.11 Noise figure comparison between state-of-the-art CMOS and SiGe LNAs above 20 GHz

from a 1.2 V supply and consumes 15.6 mW DC power. Gao et al. report a similar-performing LNA in 45nm CMOS-SOI [19]. This LNA achieves $>$ 10 dB gain over the 80 – 95 GHz band with a minimum noise figure of 4.2 dB, while the noise figure is $<$ 4.9 dB over the 85 – 95 GHz band.

5.4 Power Amplifiers

Communication systems are faced with the perpetual challenge of increasing network capacity and bandwidth. The three main leading solutions proposed are beamforming, multiple-input multiple-output (MIMO) and mm-wave carrier frequencies. These methods invariably require higher density base transceiver stations (BTS), migrating to smaller cell networks [20, 21]. Transceiver equipment used in the small-cell BTS will have to be cheap, compact and more energy-efficient than can be achieved by the current generation BTS. These factors, alongside readily available spectrum in mm-wave bands, are motivating a strong drive towards Si-based transceivers. BiCMOS transistor technology provides an excellent cost versus performance balance.

5.4.1 Millimeter-Wave PA Design Considerations

Satellite communication and wireless backhaul in the 71 – 76 GHz (uplink) and 80 – 86 GHz (downlink) bands, as well as automotive radar (77 – 78 GHz), are typical applications for E-band PAs [22–24]. While complex modulation schemes such as 64-QAM orthogonal frequency division multiplexing (OFDM) are required to achieve multi-Gb/s data rates, these schemes exhibit peak-to-average power ratios (PAPR) exceeding 9 dB, placing severe constraints on the transmitter architecture and, in particular, the PA [25]. Specifically, operating at backed-off power levels to accommodate amplitude-modulated signals severely hampers efficiency, seeing that PAs reach maximum efficiency when operating at saturated power (P_{sat}). In the literature PAs achieve around $6 - 15\% PAE$, and this drops substantially when operating at a modest back-off of 6 dB. Primarily, the large differential between P_{sat} and $OP_{1\text{dB}}$ causes the poor efficiency performance.

The antenna gain and PA output power product determine the range of a wireless transmitter. Generally, the priority of the integrated circuit (IC) designer is to maximise the power delivered to the antenna, and therefore, the power generated in the PA. Moreover, several key metrics define the performance of a PA, and obtaining a balance between these characteristics will primarily be driven by the application. The small physical size and low breakdown voltages of the high-speed transistors required to reach mm-wave ranges of transition and maximum frequencies (f_T, and f_{max}) make these devices inherently incapable of good output power. In other words, the transistor scaling applied to increase the frequency has the adverse effect of reducing the breakdown voltage and, in turn, the power handling capability of the device. There is thus a fundamental tradeoff between frequency and output power. Considering the design of PAs operating in mm-wave bands, single-transistor amplifiers cannot reach the required output power levels. Typical requirements can vary from the 10 dBm range in 60 GHz WiFi applications and extend well into the 20 dBm range for future 5G backhaul applications [26–28]. As such, additional circuit or transistor design techniques are required to raise the output power to a respectable level.

The other fundamental consideration with PAs is their role in power saving, which is significant since PA power consumption constitutes the majority of the power budget in a wireless transceiver [29]. Achieving high power-added efficiency (PAE), a metric that includes the DC current consumed by the PA, is therefore a coveted yet practically challenging design target.

5.4.1.1 Performance Characterisation

PA performance metrics can be quantified together by the first FoM (denoted by FoM_1 [30]), given by

$$FoM_1 = P_{out} \cdot PAE \cdot G_P \cdot f_c^2 \tag{5.32}$$

An improvement on this metric that accounts for the transistor f_{max} leads to

$$FoM_2 = P_{out} \cdot PAE \cdot G_P \cdot \left(\frac{f_c}{f_{max}}\right)^2 \tag{5.33}$$

These two metrics indicate that a process with higher f_{max} can be leveraged to create larger devices with better power handling (and thus, higher P_{out}) while maintaining the required power gain G_P. High output power necessitates high current consumption and large devices, sensitive to parasitics, and it is clear that P_{out} is limited by the operating frequency and process. This limitation is further complicated by the gain roll-off experienced by the device as f_c increases, instantiating a three-way balancing act between P_{out}, maximum frequency and G_P. The power gain available in the output stage is also closely tied to PAE, and it becomes increasingly challenging to reach the theoretical limit as f_c increases and V_{CC} decreases. It is therefore imperative that P_{out}, G_P and PAE are optimised for the designer to maximise FoM.

5.4.2 *State-of-the-Art Silicon PAs*

The final section of this chapter highlights performance trends in Silicon PAs. Linear and switching mode amplifiers are considered separately since the two modes incur very different design objectives. The PA survey maintained at Georgia Tech is a solid reference for monitoring PA performance trends [31].

5.4.2.1 The Prominence of Si and SiGe

The output power capabilities of Si and SiGe devices are nowhere near those of III-IV technologies. Fortunately, Si and SiGe are unmatched in terms of cost and integration capability when produced en masse. Si devices are also capable of much higher frequencies. Although InP devices can reach f_{max} of several hundred GHz, such technologies are still far too expensive and unreliable. CMOS PAs rely on gate scaling to the extreme, from the 180 nm feature size usually found in bulk CMOS down to the advanced 28 nm. In some cases, operating frequency is not the primary driving force for improving the technology. This is evident in the fact that PAs are being developed for 5G in the 28 GHz and 39 GHz bands [25, 32, 33] (relatively low frequencies in mm-wave terms) in 40 and 45 nm CMOS nodes. The main issue associated with downscaling is the reduction in breakdown voltage, and even though techniques like power combining and transistor stacking can address this, they are not without problems [34].

BiCMOS technologies that offer bipolar and field-effect devices on one substrate provide excellent flexibility. Analogue circuits can be implemented with bipolar devices, while the CMOS sections can be used for digital functions, all on one substrate. This is generally considered to be the most substantial advantage that Silicon technology offers.

5.4.2.2 Performance Trends

This section considers reported PAs in the 20 – 50 GHz and > 50 GHz range. First, Fig. 5.12 shows a compact summary of published PAs in both SiGe and CMOS technology operating in the 20 – 50 GHz range [31], which is discussed accordingly.

The performance trends highlighted by the blue (CMOS) and red (SiGe) dotted lines show oddly similar performance for both technologies. SiGe has a slight advantage in terms of P_{sat}, and more so in the lower 20 – 50 GHz range. This warrants a discussion around the best-performing amplifiers in both CMOS and SiGe. The inherent advantages of high-performance analogue bipolar devices are well documented, but looking at the trend lines in Figs. 5.12 and 5.13, the difference may not be as significant as one would expect.

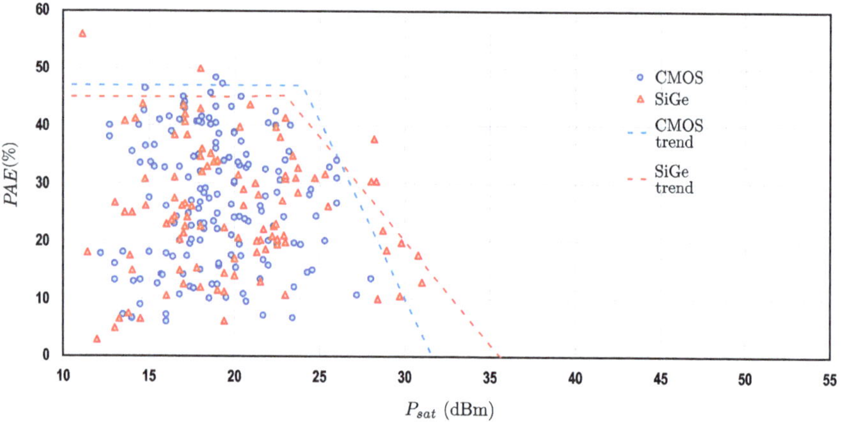

Fig. 5.12 Performance trends for SiGe and CMOS PAs between 20 and 50 GHz

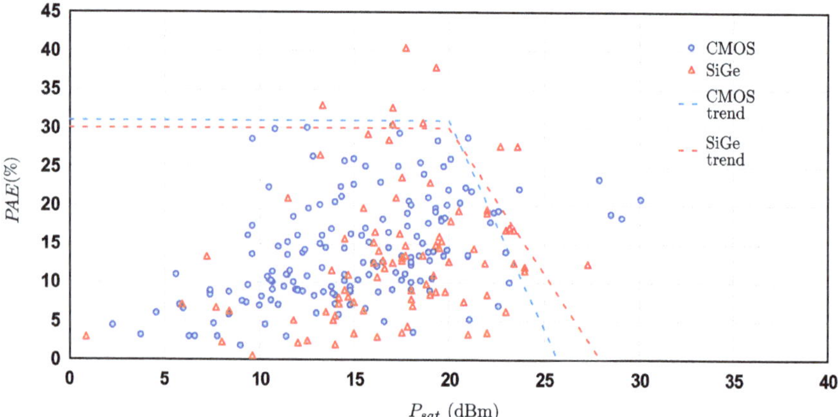

Fig. 5.13 Performance trends for SiGe and CMOS PAs above 50 GHz

Table 5.2 Wang & Wang $24-30$ GHz PA performance summary [35]

Performance metric	24 GHz	28 GHz
$PAE_{max}(\%)$	37.8	30.4[a]
P_{sat}(dBm)	28.8	28.3
P_{1dB}(dBm)	26.6	26.8
PAE(6dB backoff) (%)	27.8	21.2

[a] The PA achieves $PAE = 30.2\%$ at P_{1dB}

The discussion begins with the $20-50$ GHz amplifiers. The PA reported by Wang and Wang targets 5G new radio applications in the $24-30$ GHz band [35]. It averages over 19 dBm output power and more than 19% PAE. Furthermore, the amplifier is implemented with a Doherty architecture to boost linearity and improve efficiency at backed off input power. This is especially useful for complex modulation schemes, where amplifier efficiency drops substantially as the input power drops. Table 5.2 summarises the performance of this PA at 24 and 28 GHz.

CMOS PAs are competitive in terms of PAE, with many designs plotted in Fig. 5.12 achieving between 40 and 50%. Garay et al. report $PAE = 50\%$ for a 45 nm CMOS-SOI PA, also targeted at 5G New Radio systems operating in the 30 GHz region [36]. Their architecture utilises a dual-drive topology, which drives the gate and source terminals concurrently. Utilising the dual-drive coupled with waveform shaping techniques, this approach boosts drain efficiency and ultimately PAE.

Next, the amplifiers above 50 GHz are discussed. Figure 5.13 summarises the data for amplifiers operating above 50 GHz.

The peak PAE of 40.4% achieved by Song et al. is yet to be surpassed. This PA achieved $P_{sat} = 17.7$ dBm in an IHP 130 nm SiGe BiCMOS technology [37]. The authors implemented a cascode architecture with multi-finger transistors in parallel to boost its current-handling capability. In addition, the PA is operated in Class E, to which the high PAE can be attributed. Transmission line matching networks at the input and output ports ensure low-loss wideband matching over its $88-96$ GHz bandwidth [38, 39].

From Fig. 5.13, SiGe designs evidently have an advantage in terms of PAE, given that the nearest CMOS PA with comparable output power achieves $PAE = 29.3\%$ and $P_{sat} = 17.4$ dBm in a 40 nm bulk CMOS process [40]. The PA is operated in Class AB and utilises a transformer-based power combining scheme. Moreover, the PA can operate in low and full-power modes, where low-power operation yields $PAE = 19.6\%$ and $P_{sat} = 12.6$ dBm. Class AB operation is typically limited in SiGe BiCMOS because of the poor frequency performance of PNP transistors required to implement Class AB stages.

Another observation from Fig. 5.13 is that Watt-level output (30 dBm) above 50 GHz seems more realistic for CMOS than SiGe, even though the situation is reversed in the $20-50$ GHz range, as evident from Fig. 5.12. Nguyen et al. report a 60 GHz PA in 45 nm CMOS-SOI using distributed active transformer power combining [41]. The amplifier achieves a 2Gsym/sec symbol rate with QAM-64 modulation alongside a 20.8% peak PAE. The PA consists of 24 differential cells that can deliver between 29.1 and 30.1 dBm output power at 60 GHz. Additionally, this PA provides linear amplification of up to QAM-64 at 20.9 dBm. The downside of this amplifier is that it consumes a relatively large chip area (3×2.2 mm^2), which is to be expected with such an extensive power combining architecture. The high PAE highlights the benefit of using the SOI substrate, while similar architectures implemented in non-SOI technologies will not achieve anywhere near 20.8% due to high substrate losses.

5.4.3 *Conclusions and Discussion*

The preceding sections has highlighted some of the critical differences between CMOS and SiGe PAs and LNAs. There are good reasons for using either technology, and many factors ultimately determine the process selected for each design. For example, CMOS-SOI is advantageous for PA performance since it enables more extensive power combining schemes due to lower losses. On the other hand, SiGe PAs have an advantage in terms of operating frequency and breakdown voltage. Even though the breakdown voltage of bipolar transistors is typically higher and single devices can achieve higher output power, this can be offset by large power combining schemes that are very difficult to implement with adequate performance in BiCMOS technologies. A careful balance of cost, performance and complexity is necessary since there is a multitude of suitable technologies available for most applications.

LNA implementations in SiGe and CMOS taper off above 150 GHz, which is where most reported designs utilise GaAs or InP technologies. SiGe and CMOS offer similar performance but often take different routes to achieve it. CMOS provides a slight advantage in terms of power consumption, but only in the order of $10-20$ mW. A larger advantage on the SiGe side is in terms of gain, where SiGe LNAs provide at least 5 dB more over the entire frequency range analysed.

References

1. Marsh S (2006) Practical MMIC design. Artech House Inc., Boston, Massachusetts
2. Pozar DM (2012) Microwave engineering, 4th edn. John Wiley & Sons Inc., Hoboken, New Jersey
3. Rappaport TS, Murdock JN, Gutierrez F (2011) State of the Art in 60-GHz integrated circuits and systems for wireless communications. Proc IEEE 99(8):1390–1436. https://doi.org/10.1109/JPROC.2011.2143650
4. Hueber G, Niknejad AM (eds) (2019) Millimeter-wave circuits for 5G and radar. Cambridge University Press, Cambridge, United Kingdom
5. Božanić M, Sinha S (2018) Millimeter-wave low noise amplifiers. Springer International Publishing
6. du Preez J, Sinha S (2017) Millimeter-Wave Power Amplifiers. Springer International Publishing

7. Sokal NO, Grebennikov A (2007) Switchmode RF power amplifiers. Elsevier Inc., Oxford, UK
8. Carey E, Lindholm S (2005) Millimeter-wave integrated circuits. Springer, US
9. Yao T et al (2007) Algorithmic design of CMOS LNAs and PAs For 60-GHz radio. IEEE J Solid-State Circuits 42(5):1044–1056. https://doi.org/10.1109/JSSC.2007.894325
10. Belostotski L, Klumperink EAM (2021) Figures of merit for CMOS low-noise amplifiers and estimates for their theoretical limits. IEEE Trans Circuits Syst II: Express Briefs, vol. Early Acce. https://doi.org/10.1109/TCSII.2021.3113607
11. Belostotski L, Jagtap S (2020) Down with noise: an introduction to a low-noise amplifier survey. IEEE Solid-State Circuits Mag 12(2):23–29. https://doi.org/10.1109/6.769270
12. Belostotski L, Jagtap S (2020) Low-noise-amplifier (LNA) performance survey
13. Spasaro M, Zito D (2019) Millimeter-wave integrated silicon devices: active versus passive-the eternal struggle between good and evil: (Invited Paper). Proceedings of the International Semiconductor Conference, CAS, vol 2019, no. 829005, pp 11–20. https://doi.org/10.1109/SMICND.2019.8923669
14. Helmi SR, Chen JH, Mohammadi S (2016) High-efficiency microwave and mm-wave stacked cell CMOS SOI power amplifiers. IEEE Trans Microw Theory Tech 64(7):2025–2038. https://doi.org/10.1109/TMTT.2016.2570212
15. Yao T, Gordon M, Yau K, Yang MT, Voinigescu SP (2006) 60-GHz PA and LNA in 90-nm RF-CMOS. IEEE Radio Frequency Integrated Circuits (RFIC) Symposium, 2006, pp 1–4. https://doi.org/10.1109/RFIC.2006.1651107
16. Fritsche D, Tretter G, Carta C, Ellinger F (2015) Millimeter-wave low-noise amplifier design in 28-nm low-power digital CMOS. IEEE Trans Microw Theory Tech 63(6):1910–1922. https://doi.org/10.1109/TMTT.2015.2427794
17. Božanić M, Sinha S (2020) Millimeter-wave integrated circuits. Springer International Publishing, New York City, New York
18. Song P, Ulusoy AC, Schmid RL, Cressler JD (2014) A high gain, W-band SiGe LNA with sub-4.0 dB noise figure. In: IEEE MTT-S International Microwave Symposium Digest, pp 4–6. https://doi.org/10.1109/MWSYM.2014.6848358
19. Gao L, Ma Q, Rebeiz GM (2018) A 4.7 mW W-Band LNA with 4.2 dB NF and 12 dB gain using drain to gate feedback in 45nm CMOS RFSOI technology. In: IEEE Radio Frequency Integrated Circuits Symposium, pp 280–283. https://doi.org/10.1109/RFIC.2018.8428986
20. Rangan S, Rappaport TS, Erkip E (2014) Millimeter-wave cellular wireless networks: potentials and challenges. Proc IEEE 102(3):366–385. https://doi.org/10.1109/JPROC.2014.2299397
21. Daniels RC, Murdock JN, Rappaport TS, Heath RW () 60 GHz wireless: up close and personal. IEEE Microw Mag 11(7 SUPPL.):2010. https://doi.org/10.1109/MMM.2010.938581
22. Forstner HP et al (2008) A 77GHz 4-channel automotive radar transceiver in SiGe. Digest of Papers—IEEE Radio Frequency Integrated Circuits Symposium, pp 233–236. https://doi.org/10.1109/RFIC.2008.4561425
23. Harati P et al (2017) Is E-band satellite communication viable?: advances in modern solid-state technology open up the next frequency band for SatCom. IEEE Microwave Mag 18(7):64–76. https://doi.org/10.1109/MMM.2017.2738898
24. Pi Z, Khan F (2011) An introduction to millimeter-wave mobile broadband systems. IEEE Commun Mag 49(6):101–107. https://doi.org/10.1109/MCOM.2011.5783993
25. Chappidi CR, Wu X, Sengupta K (2018) Simultaneously broadband and back-off efficient mm-wave PAs: a multi-port network synthesis approach. IEEE J Solid-State Circuits 53(9):2543–2559. https://doi.org/10.1109/JSSC.2018.2841977
26. Xiao M et al (2017) Millimeter wave communications for future mobile networks. IEEE J Sel Areas Commun 35(9):1909–1935. https://doi.org/10.1109/JSAC.2017.2719924
27. Andrews JG, Bai T, Kulkarni MN, Alkhateeb A, Gupta AK, Heath RW (2017) Modeling and analyzing millimeter wave cellular systems. IEEE Trans Commun 65(1):403–430. https://doi.org/10.1109/TCOMM.2016.2618794
28. Taori R, Sridharan A (2014) In-band, point to multi-point, mm-Wave backhaul for 5G networks. In IEEE International Conference on Communications Workshops, pp 96–101. https://doi.org/10.1109/ICCW.2014.6881179

29. Cripps SC (2006) RF power amplifiers for wireless communications, 2nd edn. Artech House Inc., Dedham, Massachussets
30. Chen AYK, Baeyens Y, Chen YK, Lin J (2013) An 83-GHz high-gain SiGe BiCMOS power amplifier using transmission-line current-combining technique. IEEE Trans Microw Theory Tech 61(4):1557–1569. https://doi.org/10.1109/TMTT.2013.2248376
31. Wang H et al (2020) Power amplifiers performance survey 2000-present. Available: https://gems.ece.gatech.edu/PA_survey.html. Accessed 28-Feb-2021
32. Van Der Heijden MP, Scholten AJ (2018) SiGe HBT PA design for 5G (28 GHz and Beyond)—modeling and design challenges. In: IEEE BiCMOS and Compound Semiconductor Integrated Circuits and Technology Symposium, BCICTS, pp 210–214. https://doi.org/10.1109/BCICTS.2018.8551061
33. Asbeck PM (2016) Will Doherty continue to rule for 5G? IEEE MTT-S International Microwave Symposium Digest, vol 2016, pp 16–19. https://doi.org/10.1109/MWSYM.2016.7540208
34. Camarchia V, Quaglia R, Piacibello A, Nguyen DP, Wang H, Pham AV (2020) A review of technologies and design techniques of millimeter-wave power amplifiers. IEEE Trans Microw Theory Tech 68(7):2957–2983. https://doi.org/10.1109/TMTT.2020.2989792
35. Wang F, Wang H (2020) A 24-to-30GHz Watt-level broadband linear Doherty power amplifier with multi-primary distributed-active-transformer power-combining supporting 5G NR FR2 64-QAM with >19dBm average Pout and >19% average PAE. In: Digest of Technical Papers—IEEE International Solid-State Circuits Conference, pp 362–364. https://doi.org/10.1109/ISSCC19947.2020.9063146
36. Garay EF, Munzer DJ, Wang H (2021) 26.3 A mm-Wave power amplifier for 5G communication using a dual-drive topology exhibiting a maximum PAE of 50% and maximum de of 60% at 30GHz. In IEEE International Solid-State Circuits Conference, pp 258–260. https://doi.org/10.1109/ISSCC42613.2021.9365830
37. Song P et al (2015) A class-E tuned W-Band SiGe power amplifier with 40.4% power-added efficiency at 93 GHz. IEEE Microwave Wirel Compon Lett 25(10):663–665. https://doi.org/10.1109/LMWC.2015.2463231
38. Mader TB, Popović ZB (1995) The transmission-line high-efficiency class-E amplifier. IEEE Microw Guid Wave Lett 5(9):290–292. https://doi.org/10.1109/75.410401
39. Wilkinson AJ, Everard JKA (2001) Transmission-line load-network topology for class-E power amplifiers. IEEE Trans Microw Theory Tech 49(6):1202–1210. https://doi.org/10.1109/22.925525
40. Zhao D, Reynaert P (2013) A 60-GHz dual-mode class ab power amplifier in 40-nm CMOS. IEEE J Solid-State Circuits 48(10):2323–2337. https://doi.org/10.1109/JSSC.2013.2275662
41. Nguyen HT, Jung D, Wang H (2019) A 60 GHz CMOS power amplifier with Cascaded asymmetric distributed-active-transformer achieving Watt-level peak output power with 20.8% PAE and supporting 2Gsym/s 64-QAM modulation. In: International Solid- State Circuits Conference, pp 90–92

Chapter 6
Frequency Synthesis and Conversion Circuits in Millimeter-Wave Silicon

Frequency synthesis and conversion are ever-present in RF, microwave and mm-wave circuits, and they are staple building blocks in transceivers (see the familiar block diagram in Fig. 6.1).

Mixers are tasked with converting between the low intermediate frequency (IF) and mm-wave carrier frequency. The emergence of high-performance FETs has caused mixer design to move away from the traditional Schotty diode and waveguide approach. Additionally, mixers with good mm-wave performance are essential for the commercial expansion of products developed for the mm-wave range. Multiple mixer topologies are available to designers, depending on the technology and the application requirements, since the design is strongly affected by the transmitter and receiver architectures. Performance metrics typically used are conversion loss, isolation, dynamic range, compression point and intermodulation distortion.

Mixers, however, rely on periodic signals at their inputs, which is where oscillators come into play. An oscillator design must meet two primary specifications: the system specifications (e.g. output frequency) and the interface specifications (e.g. output swing and matching). Voltage-controlled oscillators are the most common application, where a tuning voltage changes the output frequency. Oscillators are also used as reference inputs to frequency synthesisers, which provide a more sophisticated method of generating frequency references throughout transceiver systems. In fact, wherever integrated transistors have gain (as limited by frequency), oscillators can be realised. MMIC oscillators are limited to weak output power, primarily as a result of their thermal breakdown characteristics. As such, further amplification is almost always necessary.

J. du Preez and S. Sinha, *State-of-the-Art of Millimeter-Wave Silicon Technology*, Lecture Notes in Electrical Engineering 945,
https://doi.org/10.1007/978-3-031-14655-8_6

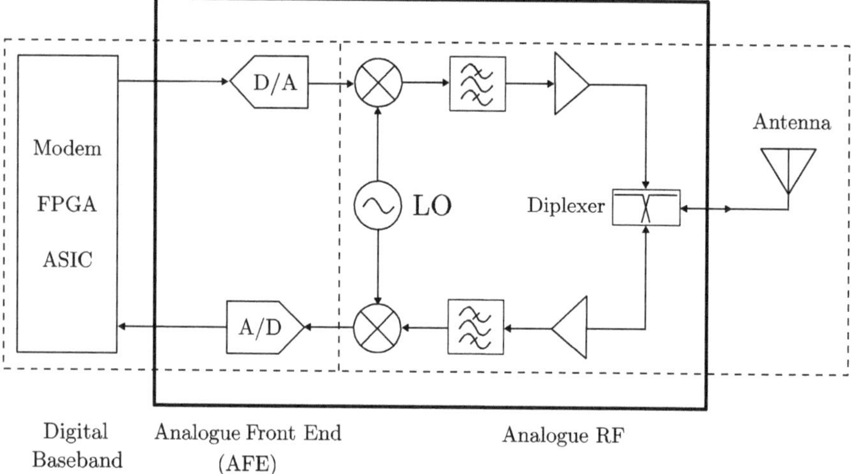

Fig. 6.1 RF transceiver block diagram

6.1 Oscillators

This section provides an overview of oscillator principles, and additionally, highlights challenges with mm-wave design alongside some of the approaches followed in state-of-the-art designs. Signal synthesis is broadly divided into direct or indirect methods, and the distinction is based on whether the LO output frequency is the same or higher than the fundamental frequency [1].

Integrated transistor oscillators provide less output power and can achieve lower frequency than passive, diode-based implementations. With that said, transistors provide some notable advantages. First of all, transistors are readily available for integration with surrounding circuits, most of which are also transistor-based, while diode devices are less flexible in this sense. The fabrication of the diode device must be very accurate in order for it to achieve the desired physical characteristics, which, in turn, influences its negative resistance oscillation. Transistors, on the other hand, offer great flexibility through a wide range of viable bias points as well as source and load impedance tuning. Typically, transistor oscillators offer better control over temperature stability, noise, and frequency than diodes. Moreover, improved architectures that provide frequency tuning or injection locking can easily be implemented with transistor circuits.

Tunable frequency sources are found extensively in frequency-hopping communication and radar systems and many electronic warfare applications. Voltage-controlled oscillators (VCOs) can be implemented by placing a reverse-biased varactor in the tank of an existing transistor oscillator.

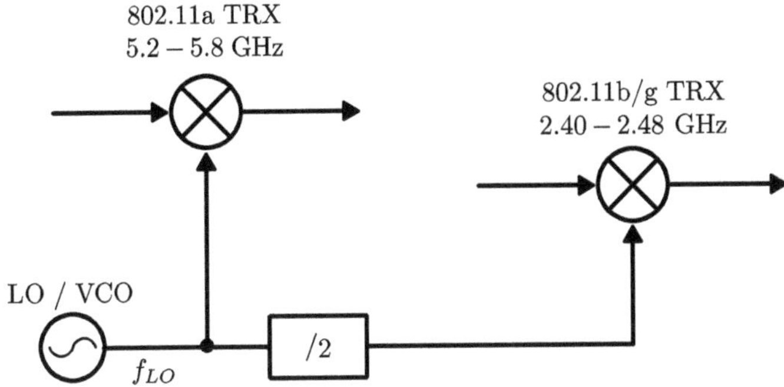

Fig. 6.2 Oscillator path for a dual-band application

6.1.1 Oscillator Performance Metrics

6.1.1.1 Tuning Range

Oscillators must typically offer at least a modest degree of tunability. For example, 900 MHz GSM systems with direct-conversion receivers may be required to tune between 935 and 960 MHz. Additionally, oscillators used in performance-critical systems incorporate different forms of temperature compensation techniques to offset induced errors. The output frequency is also dependent on whether quadrature outputs are needed, such as in direct-conversion receivers.

An example configuration is shown in Fig. 6.2, where the local oscillator (LO)[1] signal is split between two paths for an 802.11 wireless transceiver.

The LO output frequency, f_{LO} is divided by two to serve the 2.4 GHz path of the transceiver. This 802.11 LO is required to provide quadrature outputs since radios in this class use QAM modulation schemes.

6.1.1.2 Voltage Swing

Controlling the output voltage swing is an important consideration that depends on the stages connected to the LO. The LO must provide a voltage swing that is large enough to drive frequency dividers and mixer blocks. Moreover, insufficient voltage swing will worsen the effects of the oscillator's internal noise [2].

6.1.1.3 Output Drive

Oscillators typically need to drive reasonably large load capacitances, purely because there are multiple circuit blocks connected at its output port.

---

[1] A LO or a voltage-controlled oscillator (VCO) is typically used here.

6.1.1.4 Phase Noise

Practical oscillators do not have single-impulse spectra; the energy is usually concentrated around the centre frequency, but the impulse widens due to noise components. This phenomenon is known as phase noise, and it affects transceiver performance dramatically. Designers must carefully balance phase noise with tuning range and power dissipation, leading to a reasonably challenging design process. The phase noise in LC oscillators, for example, depends strongly on the tank circuit's Q-factor. This makes the mm-wave realisation problematic, especially in Silicon, since the achievable passive Q-factors are usually quite low [3].

On the other hand, amplitude noise is rejected naturally by the oscillator. This stems from the fact that the loop gain of the oscillator is unity at its desired operating frequency. The amplitude adjusts through the compression characteristic of the non-linear components and the aforementioned unity loop gain. Oscillator phase is essentially free-running, and there is no corrective mechanism that restores phase errors.

6.1.1.5 DC Supply Sensitivity and Power Dissipation

Most oscillators exhibit some level of sensitivity to variations in the DC supply voltage, an undesirable characteristic that translates into frequency and phase noise. Low-frequency noise generated in power circuits is especially difficult to get rid of and exerts a modulating effect on the output frequency.

The power dissipated in the LO subsystem is crucial in some applications. Typically, techniques to minimise phase noise for a given tolerable power dissipation are implemented.

6.1.2 Basic Oscillator Operating Principles

Circuit oscillation requires (1) unity gain and (2) 360° phase shift. Practically speaking, this problem is essentially non-linear and must be solved iteratively since the active device should have *some* gain greater than unity at start-up to initiate oscillation. Without this, circuit losses and noise effects will prevent the oscillator from starting up at all. When the oscillation stabilises around the desired setpoint, the operating point of the active device has shifted to where the net gain is unity. The second condition (360° phase shift) is satisfied when the net reactance of the circuit is zero.

Microwave oscillators consist of a resonator, gain and load. The resonator sets the centre frequency and ensures the 360° phase shift, and the gain block generates power, typically via a positive-feedback transistor. The simplest resonators are series LC circuits, which resonate at a frequency $f = 1/2\pi\sqrt{LC}$ and induce a net zero phase shift. Moreover, the load that the oscillator sees is the system impedance or the impedance of whichever block is connected to its output. Modelling the oscillator in this way simplifies simulation since the components can now be separated and simulated with linear models to verify phase characteristics.

6.1.2.1 Resonators

The resonator is the core component of almost all integrated oscillators. These resonators are almost exclusively implemented with different combinations of capacitors and inductors. As such, the achievable Q-factor is low. Since the resonator consists of two reactive components, it is a second-order system that will oscillate at a frequency $f_0 = 1/2\pi\sqrt{LC}$ when positive feedback is introduced. However, due to the damped nature of the LC resonator, the oscillating waveform will gradually decay to zero. As such, the damping effect must be eliminated for the oscillation to remain. Practical oscillators utilise feedback in order to sustain oscillatory behaviour.

6.1.2.2 Negative Resistance Circuits

Oscillators require some feedback circuit to sustain oscillation and prevent decay that results from damped behaviour. A negative resistance feedback loop, as Fig. 6.3 shows, can generate a negative resistance.

Negative resistance can be realised via a feedback circuit. Using a tapped capacitor coupled with an amplifier leads to the Colpitts architecture, shown in Fig. 6.4a. Conversely, using a tapped inductor coupled with an amplifier leads to the Hartley architecture, as Fig. 6.4b illustrates. Due to the complex layout of integrated inductors, Hartley oscillators are seldom implemented.

6.1.3 Oscillator Architectures

Direct LO synthesis is challenging in mm-wave bands and poses severe difficulties in achieving the desired phase noise and tuning range requirements. Pushing the fundamental frequency f_0 closer to the transistor f_{max} causes a reduction in the gain that can be extracted from the device. As such, the power must be ramped up

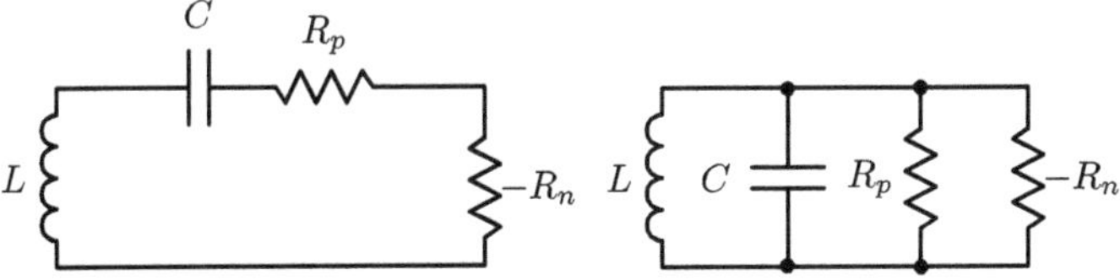

Fig. 6.3 Negative resistance added to series and parallel LC resonators

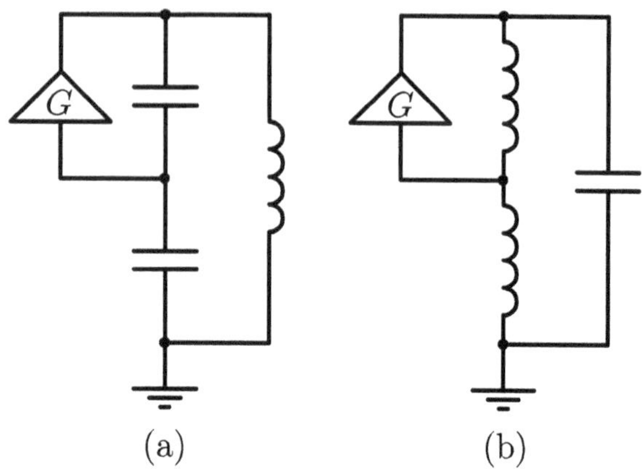

Fig. 6.4 Practical feedback resonators. **a** Colpitts oscillator **b** Hartley oscillator

to sustain oscillatory behaviour. Additionally, the Q-factor of capacitors, inductors and varactors fabricated in lossy silicon substrates degrades substantially at higher frequencies [4]. This affects the start-up power required and diminishes phase noise performance.

The parasitic components of the interconnects, oscillator core and buffer stages are significant at mm-wave frequencies and strongly influence circuit behaviour and performance. The frequency tuning range is limited due to circuit parasitics, and only a relatively small MOS varactor may be used. Furthermore, parasitic component values become close to the computed component values needed to implement a paper design.

6.1.3.1 Voltage-Controlled Oscillators

The output frequency of a voltage-controlled oscillator (VCO) can, as the name implies, be tuned by a reference voltage. A typical application is phase-locked loops (PLLs), where the VCO fulfils the carrier recovery function. Varactors are most often used to implement frequency control. MOS varactors are used in BiCMOS processes, while bipolar technologies can utilise base–collector or base–emitter junctions. As with any junction, applying a potential across the junction gives rise to parasitic capacitance. Base-emitter and base–collector junction capacitances have reasonably high Q, unlike the MOS varactor, where the Q is reduced due to low substrate doping. Varactor placement is crucial since it plays a significant role in low-frequency noise. Moreover, MOS varactors require both positive and negative voltages, which influences the layout and design of the biasing circuit.

6.1.3.2 Transistor Oscillators

Figure 6.5 shows the circuit for a simple bipolar Colpitts oscillator.

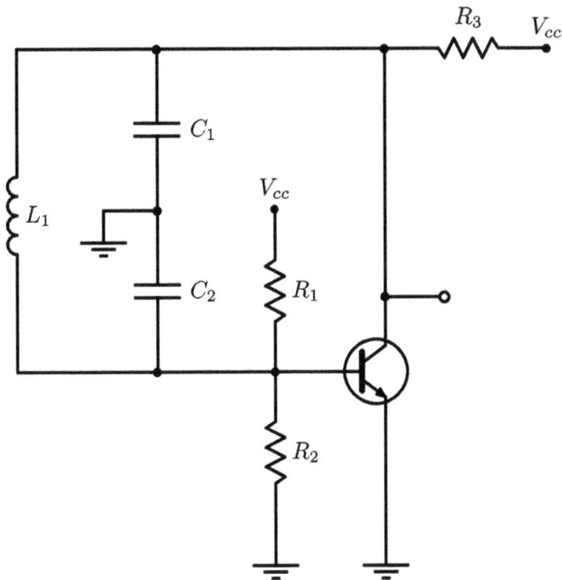

Fig. 6.5 Bipolar Colpitts oscillator

This configuration uses two centre-tapped capacitors with a shunt inductor to form its resonant tank circuit. The Colpitts configuration also provides good frequency stability and is known for good phase noise characteristics [5]. HBTs used in integrated Colpitts designs are biased at their minimum noise figure current density point (J_{opt}).

As shown in Fig. 6.6, the Hartley oscillator is essentially the opposite of the Colpitts circuit. This configuration utilises two centre-tapped inductors with a shunted capacitor to form its tank circuit.

Figure 6.6 shows the circuit for a simple bipolar Hartley oscillator.

The fact that the Hartley architecture needs two inductors is detrimental to mm-wave implementation, where inductors with high Q-factor are difficult to synthesise and are costly in terms of routing and chip area.

6.1.3.3 Frequency and Power Scaling

Two general rules describe oscillator frequency and power scaling operating in mm-wave bands [6]. First, at some frequency f_0, the oscillator phase noise can be reduced by a factor $10 \log N$ by tuning the transistor size, current consumption and passive admittance values by N. This is independent of the topology. Second, for a given power consumption, the phase noise of an oscillator with fundamental frequency at $M f_0$, which was created by passive scaling an f_0-oscillator, scales by $20 \log M$. An example of power scaling is shown in Fig. 6.7, where a cross-coupled LC oscillator is scaled by a factor N. Transistor emitter length (L_E) and bias current (I_E) are scaled along with the passive resonator components.

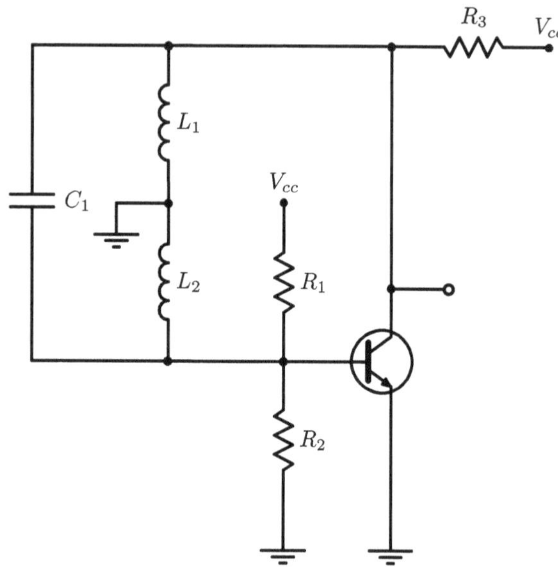

Fig. 6.6 Bipolar Hartley oscillator

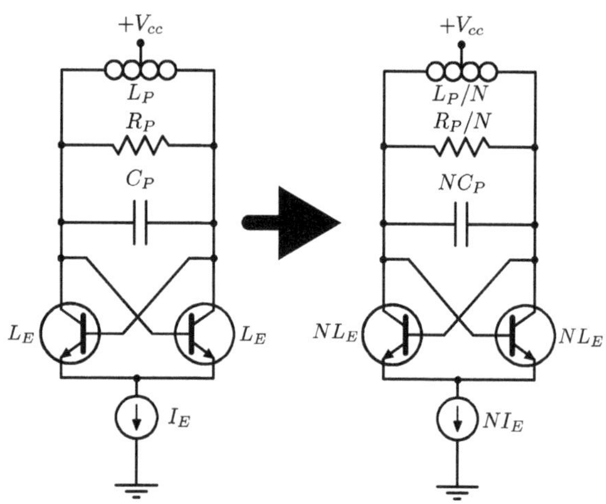

Fig. 6.7 Scaling a cross-coupled LC oscillator

Large values of N will render the passive component values impractical. For example, for a 73 GHz oscillator to achieve −85 dBc/Hz phase noise at a 1 MHz offset, the inductor must be 1 pH, which is not practical [6]. This example uses a 130 nm SiGe BiCMOS technology with f_{max} close to the oscillation frequency. As such, the phase noise is already limited from the outset. One possible solution is to couple multiple oscillators.

Frequency scaling of a conceptual Colpitts oscillator is illustrated in Fig. 6.8.

The frequency scaling condition described earlier in this section holds true only if the passive Q-factor remains consistent with scaling and the transistor parasitics

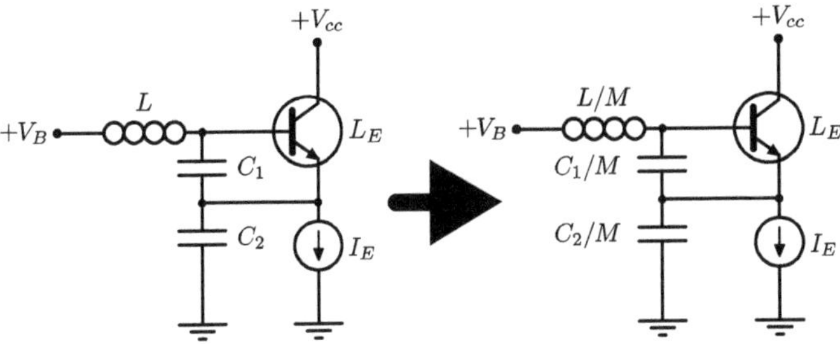

Fig. 6.8 Frequency scaled Colpitts oscillator

can be neglected. As the oscillation frequency approaches the transistor f_{max}, these conditions will no longer be valid and must be analysed in greater detail.

Phase Noise in Scaled Mm-wave Oscillators

The circuits in Fig. 6.9 illustrate a basic mm-wave Colpitts oscillator in (a), and its large-signal equivalent in (b).

The collector current is modelled as

$$I_C = I_s e^{V_\pi / m V_T} \tag{6.1}$$

where:

- I_s is the saturation current.
- m is a technology parameter.
- V_T is the thermal voltage, equalling 26 mV at room temperature.

The transistor f_{max} is largely a function of its parasitics, and is determined by

$$f_{max} = \frac{1}{4\pi}\sqrt{\frac{g_m}{(C_\pi + C_\mu)C_\mu R_B}} \tag{6.2}$$

6.1.3.4 Sub-Harmonic Injection Locking

Phase-locked oscillators require a number of components: VCOs, frequency dividers and multiple phase and frequency comparators. As a result, the oscillator starts occupying a notable chip area and consumes non-negligible power. Multi-chip packaging may also be required, further driving up costs. The limitations of mm-wave dividers mean that additional multipliers and amplifiers must be implemented at mm-wave. The multipliers, in turn, have stringent filtering requirements at their output ports.

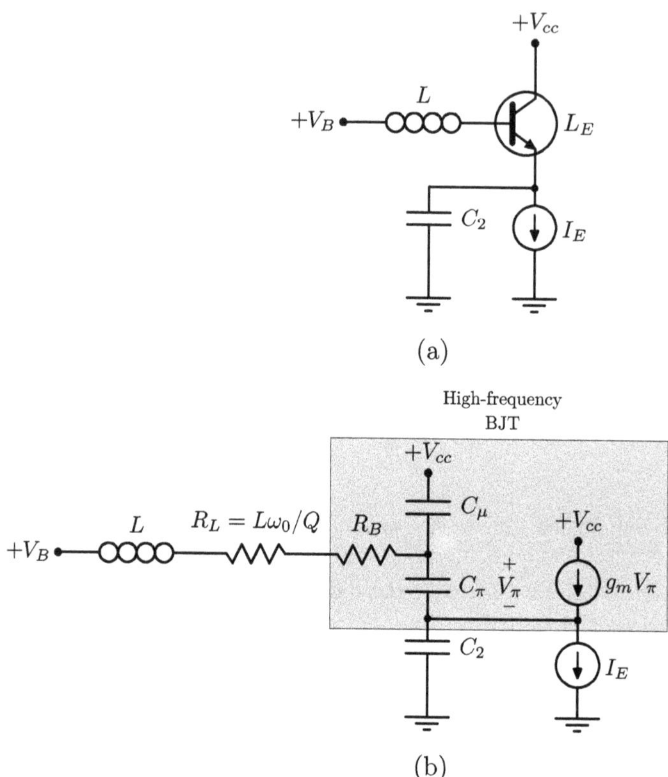

Fig. 6.9 **a** Mm-wave Colpitts oscillator circuit. **b** High-frequency equivalent large-signal model

All these factors make it highly challenging to implement integrated synthesisers. Techniques typically used in VCO design such as segmentation with a varactor or switched capacitor are not effective when operating close to f_{max} [7]. A VCO with large gain (K_{VCO}) used within a PLL loop introduces up-converted noise originating from the charge pump and loop filters.

A viable alternative is distributing a low-frequency reference and injecting it into mm-wave oscillator blocks. The oscillators then lock on to some harmonic of the injected signal [8–11]. In essence, the oscillators operate as frequency multipliers.

This type of synthesiser, which consists of a sub-harmonic PLL followed by a frequency multiplier, enables lower phase noise and higher tuning range. Frequency multiplication causes an increase in in-band noise of $20\log M$; f_0^2 term in Leeson's equation, which describes single-sideband phase noise,

$$L(f_m) = 10\log\left[\frac{1}{2}\left(\left(\frac{f_0}{2Q_{1f_m}}\right)^2 + 1\right)\left(\frac{f_c}{f_m}\right)\left(\frac{FkT}{P_s}\right)\right](\text{dBc/Hz}) \qquad (6.3)$$

where:

- f_0 is the output frequency
- Q_1 is the loaded Q-factor
- f_m is the frequency offset, relative to f_0
- f_c is the $1/f$ corner frequency
- F is the amplifier noise factor
- k is Boltzmann's constant
- T is the absolute temperature
- P_s is the power available at the amplifier input.

As a result, the improvements in Q, voltage swing at the output and tuning range are retained in addition to improvement in the phase noise [7].

An example application of a sub-harmonic injection-locked synthesiser is shown in Fig. 6.10, where the PLL output frequency is tripled before being fed to the TX and RX quadrature mixers.

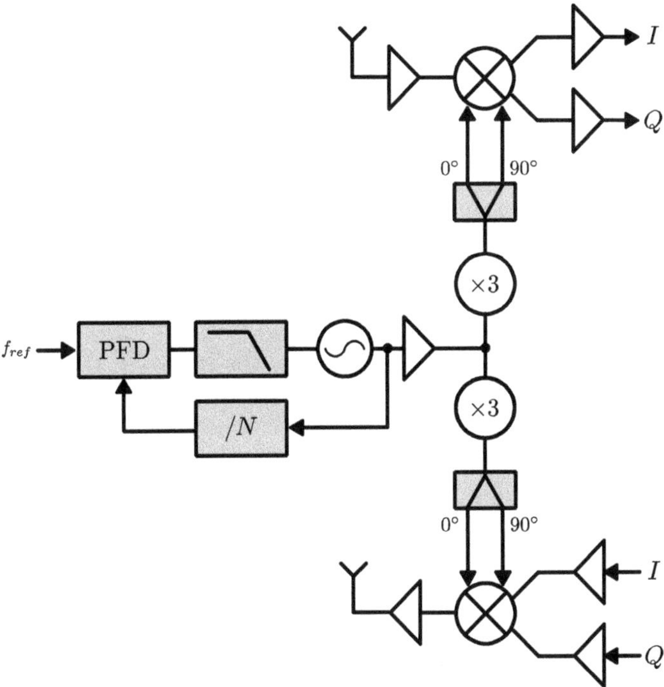

Fig. 6.10 Direct-conversion transceiver implementation with a sub-harmonic injection-locked synthesiser

6.2 Mixers

Mixers are three-port components used to implement frequency conversion through non-linear elements. Ideally, a mixer produces an output signal consisting of the sum and difference frequency components of its two inputs. Non-linear components are known to produce a number of harmonics, and a filter is implemented to select the correct output frequency. Mixers are used extensively in wireless transceivers, both in analogue and digital form.

6.2.1 Basic Mixer Operating Principles

Figure 6.11 shows both an up- and down-conversion mixer. The up-conversion mixer accepts a reference LO and a signal at baseband or some intermediate frequency (IF) and produces an RF output. Conversely, the down-conversion mixer produces f_{IF} given f_{LO} and f_{RF}.

From a reference LO signal,

$$v_{LO}(t) = \cos(2\pi f_{LO} t) \tag{6.4}$$

and an IF signal

$$v_{IF}(t) = \cos(2\pi f_{IF} t) \tag{6.5}$$

the output of the up-conversion mixer, ideally, is

$$v_{RF}(t) = \cos 2\pi(f_{LO} - f_{IF})t + \cos 2\pi(f_{LO} + f_{IF})t \tag{6.6}$$

The sum and difference components adjacent to the carrier frequency are the upper and lower sidebands. Some architectures, such as complex mixers, will produce

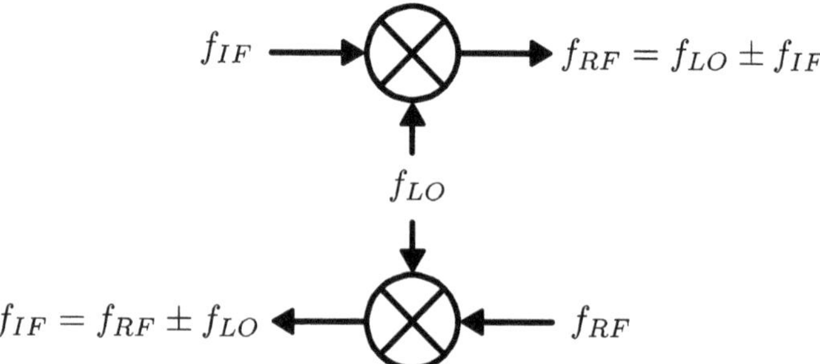

Fig. 6.11 Block diagram of a mixer in up and down-conversion operation

single-sideband (SSB) outputs. The resulting IF signal following down-conversion can be obtained by multiplying v_{LO} with v_{RF}. Receivers will use a lowpass filter to separate the $f_{IF} = f_{RF} - f_{LO}$ component for digitisation and further processing. Practically, the mixer outputs will contain harmonics since the multiplication is achieved with non-linear devices. Additional filtering is generally added for harmonic suppression.

6.2.1.1 Image Frequency

The receive antenna in a wireless system typically has a wider bandwidth than the band of interest.

6.2.1.2 Noise and Noise Figure

The frequency translation that takes place within the mixer complicates the noise figure characterisation. The noise factor in a mixer circuit is

$$F = \frac{N_{0,tot}(\omega_{IF})}{N_{0,source}(\omega_{IF})} \tag{6.7}$$

where the two noise components are the total output noise at IF (numerator) and the output IF noise due to the source (denominator). The circuit produces noise at all frequencies, and at least some of the noise will appear at the output IF as a result of the mixing action. The input and IF components usually dominate.

Mixer specification sheets also list single and double sideband noise figures. In the latter case, the denominator in (6.7) contains all the noise at the output IF, which includes noise due to the image frequency. On the other hand, the single-sideband mixer noise factor calculation only uses output RF noise. The ideal single-sideband mixer can achieve no better than 3 dB noise since the source noise doubles due to mixing the RF and IF noise into the IF.

Minimising the mixer noise figure is generally not a concern, as it is placed after the LNA in most receivers. However, low-gain mm-wave LNAs may require a lower mixer noise figure to achieve a lower overall noise figure.

6.2.1.3 Linearity

The non-linear components used to implement mixers produce intermodulation components. As such, mixers generally specify IIP_3 in the 15 to 30 dBm range. Distinguishing between mixer linearity and non-linearity is often confusing. On the one hand, mixers must operate as non-linear devices to realise their mixing characteristic. On the other hand, to produce high-quality conversion, mixer performance must be linear. In other words, they operate non-linearly but perform linearly. The maximum tolerable input signal is determined by specifying the 1 dB compression point, and the maximum level of converted RF signals that end up near the IF is characterised by the intermodulation products.

1-dB Compression Point

Like amplifiers, mixers *perform* linearly when the RF input is small enough for the conversion gain to remain consistent. As the RF input power increases, the mixer will eventually saturate, compressing the mixer. The 1 dB compression point measures the departure from the linear characteristic. The input power required to drive the mixer into compression is

$$P_{1\mathrm{dB}}^{in} = P_{1\mathrm{dB}}^{out} - G - 1\ \mathrm{dB} \tag{6.8}$$

where G is the conversion gain.

Third-Order Intercept

Mixing of the RF and LO (and their respective harmonics) inevitably causes intermodulation products to appear at the mixer output. In multi-channel receivers, multiple closely spaced RF components are possible and must be contended with.

6.2.1.4 Isolation

Isolation is another critical characteristic since LO leakage into the RF port can be detrimental to performance. LO power coupled to the antenna can cause spurious emissions and cause the system to fail qualification, seeing that it will likely interfere with adjacent users or channels. The isolation is primarily influenced by the coupler architecture used to diplex the RF and LO ports.

6.2.2 Mixer Architectures

Mixers are generally categorised as active or passive components, depending on which non-linear component is used to realise the mixing operation. Active transistor mixers use devices biased in such a way that the transistor provides gain, which means that the mixer can provide conversion gain (as opposed to loss). However, active mixers cannot be reciprocal, so the same design cannot be utilised for both up and down-conversion. Passive mixers are most often implemented with diodes, but unbiased transistors may also be used. Routing DC bias to diode mixers is generally easier.

6.2.2.1 Single-Ended Mixers

A single-ended mixer is constructed with one non-linear device, a matching network for the IF port, and a matching-combiner network for the RF and LO ports. Figure 6.12 shows a single-ended mixer.

While Fig. 6.12 shows a series diode as the non-linear component, shunted diodes are also often used. Moreover, the diode can be biased or not, depending on the required LO power levels. Single-ended mixers suffer from a large number of harmonics present at their output.

Using a three-terminal device as the non-linear element is also possible. Figure 6.13 shows such a configuration.

When no drain bias is applied to the transistor, the mixer is operating in its passive mode. This configuration can be advantageous since this mixer is reciprocal, so that it may be used for up or down-conversion. The RF input is connected to the source, and the LO source is connected to the gate terminal. As shown in Fig. 6.13, the IF output is taken at the drain terminal.

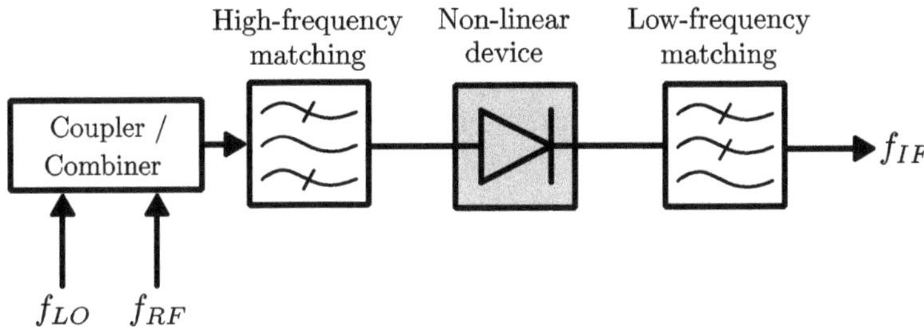

Fig. 6.12 Single-ended mixer

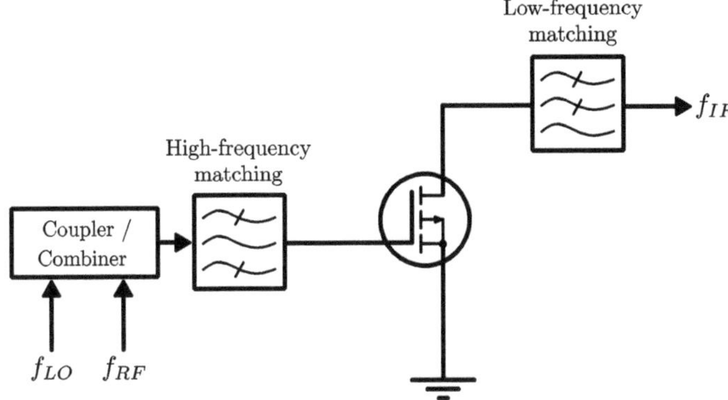

Fig. 6.13 Single-ended mixer with a FET as the non-linear device

An alternative to the configuration shown in Fig. 6.13 is to use a dual-gate FET. This eliminates the requirement for a combiner or coupler at the RF/LO ports of the mixer. Additionally, this arrangement provides naturally good isolation.

6.2.2.2 Balanced Mixers

Figure 6.14 shows a balanced mixer. It uses two non-linear elements, each driven with 180° out of phase RF and LO signals. The phase offset is realised with a 3 dB splitter, and the diode outputs are again combined in-phase.

As is the case with single-ended mixers, the diodes may be connected in series or parallel, with or without a DC bias voltage applied. Harmonic suppression of the even LO harmonics can be achieved by using a 180° coupler at the input ports. The circuit in Fig. 6.14 may also be implemented with a 90° hybrid.

6.2.2.3 Double-Balanced Mixers

Double-balanced mixers are an extension of the single-balanced architecture and, in fact, merely consist of two single-balanced mixers connected in parallel with a 180° phase shift between them. This architecture provides even harmonic suppression for both RF and LO frequencies (as opposed to only LO components). However, the RF and LO ports require baluns, as Fig. 6.15 shows, complicating the integration process.

6.2.2.4 Image-Reject Mixers

A single-ended mixer, as shown earlier, produced sum and difference outputs of the IF and LO signals (in the up-conversion case). In most cases, only one of these components is desirable, be it the upper or lower sideband. The undesired frequency is called the image frequency, and it adds unwanted noise to the output of the mixer.

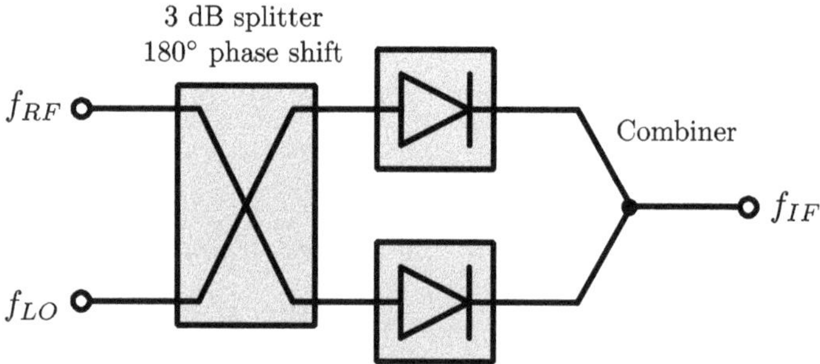

Fig. 6.14 Balanced mixer

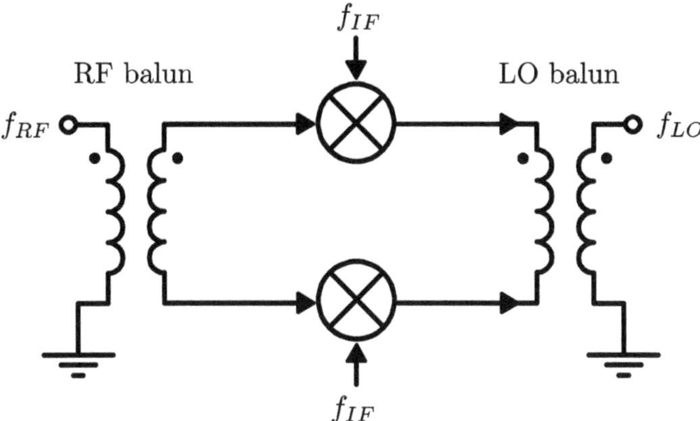

Fig. 6.15 Double-balanced mixer

The image-reject mixer, shown in Fig. 6.16, eliminates the image components from the IF output port, which can improve the noise figure by up to 3 dB.

The image-reject mixer uses two double-balanced mixers (drawn as single mixers in Fig. 6.16) where a 90° hybrid feeds the RF path, and an in-phase splitter network feeds the LO path. The upper and lower sidebands are separated at the output IF port through another 90° hybrid.

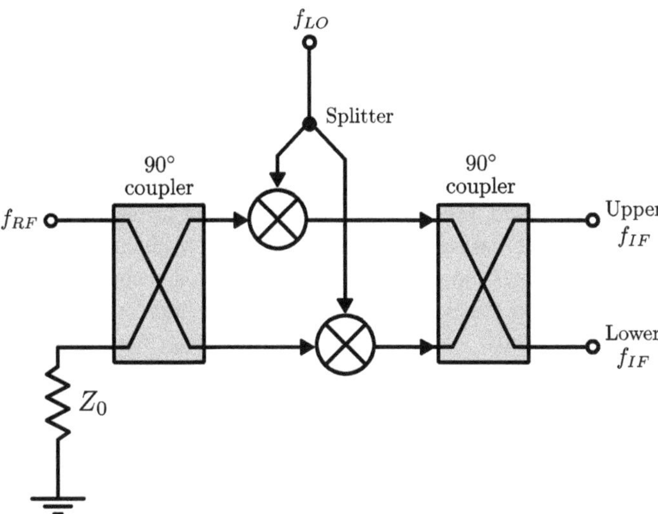

Fig. 6.16 Image-reject mixer

6.2.3 Mm-wave Mixer Design

Next-generation communication systems such as 5G new radio (NR), radar and high-speed backhaul are targeted at high mm-wave bands [12–14]. These bands, which include the V-band (40 – 75GHz), E-band (71 – 76 GHz, 81 – 86 GHz and 92 – 95 GHz) and W-band (75 – 110 GHz), are good contenders for massive MIMO and phased array systems [13, 15–17]. Mixers are at the forefront of mm-wave transceivers, and most transceivers make use of mixer-first or LNA-first receive paths.

6.2.3.1 Mixer-First Receivers

The LNA-first architecture mentioned in the preceding section provides the obvious benefit of a lower noise figure at the cost of linearity and selectivity. In a crowded RF environment, selectivity is critical in order to prevent out-of-band interference and subsequent saturation of the receiver [18]. This is where the mixer-first architecture is advantageous. For one, it can provide much wider frequency coverage by virtue of its first-stage bandpass filtering and wide tuning range. Additionally, this architecture generally performs better in terms of linearity, offering improved intermodulation performance. This is a crucial characteristic considering massive MIMO, as these systems are inherently sensitive to intermodulation and cross-modulation.

Regardless of the lower noise figure (compared to the LNA-first receiver), mixer-first receivers benefit from SNR that is further improved by the array gain in the beamformer [19]. Figure 6.17 shows a block diagram of a mixer-first beamforming receiver.

The receiver shown in Fig. 6.17 can be used, as an example, in multi-user MIMO applications [20, 21]. The massive MIMO approach provides noise averaging, where M access point transceivers enhance the SNR by $10 \log M$, relative to a single-element receiver. As such, the noise figure requirement for each individual receive channel is relaxed. Increasing the number of receiver channels increases power consumption and requires more chip area, each of which is challenging to solve.

LNA-first mm-wave receivers require reasonably high power consumption to achieve gain above 20 dB, mainly because multiple stages are needed. The added effect of this is that interstage matching networks are required, further increasing the footprint size. Of course, the number of stages is fundamentally limited due to the additive parasitic elements and, once again, limited chip area. One can imagine that large MIMO arrays are incompatible with this architecture, which is why many authors suggest mixer-first implementations [19, 22–24].

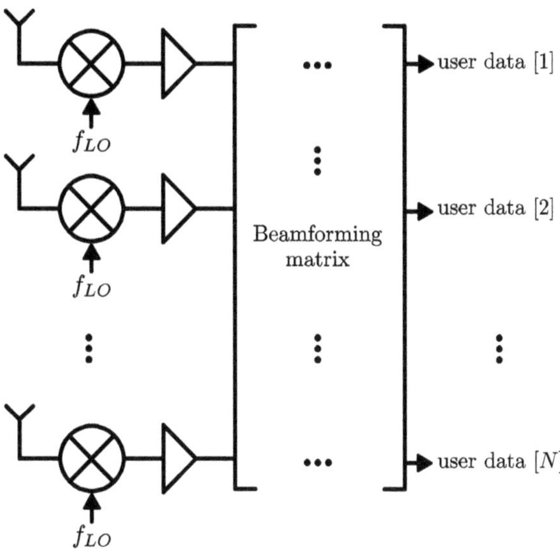

Fig. 6.17 Mixer-first beamforming receiver

6.2.3.2 N-Path Filters

Implementing tunable wideband RF filters is challenging. State-of-the-art mm-wave on-chip filters typically offer simultaneous tuning of centre frequency and bandwidth but rarely achieve more than an octave of bandwidth. A popular alternative is to use N-path filters, where the centre frequency is tunable by shifting the LO frequency, and the bandwidth is tuned by altering the baseband impedance [25]. Both frequency and bandwidth are instantly reconfigurable. Figure 6.18 shows the circuit diagram of a conventional N-path filter.

The LTI equivalent of this filter is shown in Fig. 6.19, where $v_x = v_{in}/2$ and $\gamma = 2/\pi^2$ [23].

The basic operation of an N-path filter is as follows. Passive switches are used to perform down-conversion. Charge storage at baseband facilitates up-conversion on the next switching cycle, and each switch has a duty cycle of $1/N$. The attainable Q-factor of the N-path filter is decoupled from its passive component values, and as such, it can reach much higher values.

A key property of the N-path filter architecture is its impedance transparency, which facilitates its flexibility. The MOS devices shown in Fig. 6.18 realise simultaneous up-conversion of the baseband voltage to RF and down-conversion of the RF current back to baseband. HBTs generally give better f_T, while MOS transistors are essentially bidirectional in their triode region, which yields two-way conversion. Ying and Molnar [25] provides a thorough comparison of HBT and MOS-based N-path mixers, a summary of which is provided here.

MOSFET switching behaviour is illustrated in Fig. 6.20.

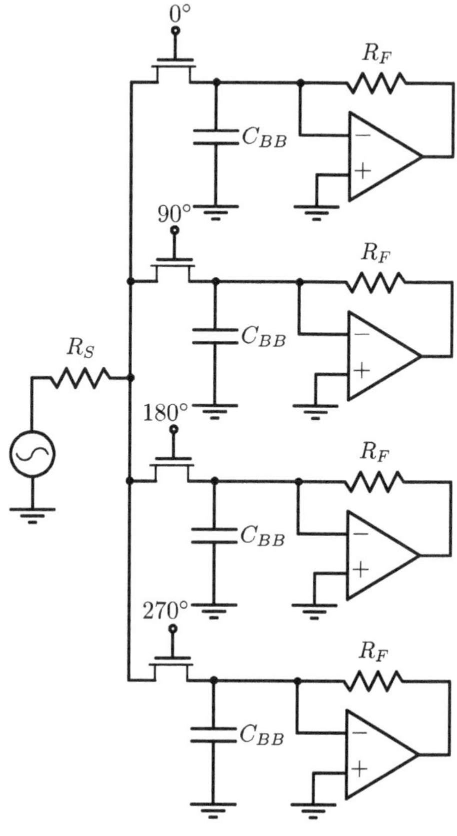

Fig. 6.18 N-path filter

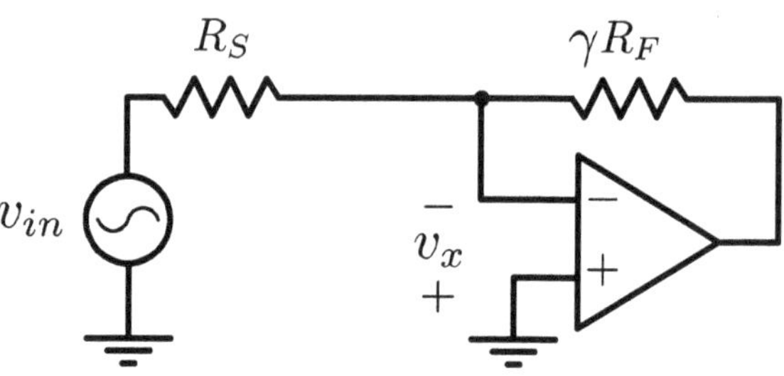

Fig. 6.19 LTI equivalent circuit of the N-path filter

Typical N-path MOSFET receivers use resonant structures for generating their LO, and often utilise advanced short-channel devices [23]. LO generation for mm-wave N-path filters is challenging due to the large voltage swing required, the non-overlapping nature of quadrature pulses, and the high logic switching speeds. First, maximising linearity requires driving the MOS devices at their maximum V_{GS}. This also reduces the switching turn-on resistance, which, in turn, affects noise figure and also linearity [26]. Second, at the overlap point where the 0° FET presents

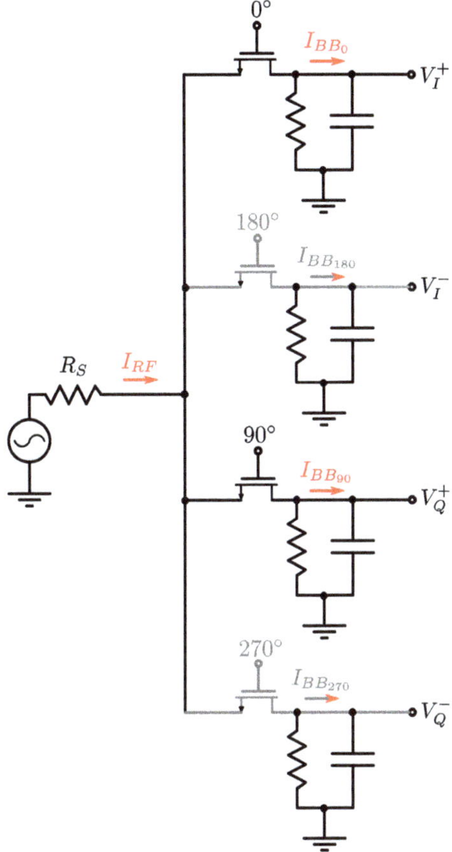

Fig. 6.20 MOSFET switching behaviour illustration. Devices that are conducting are shown in black, and non-conducting devices are in grey. Adapted from [25]

low impedance, the impedance presented by the 90° FET reduces quickly due to the switching nature of the filter arrangement. During this overlap period, adjacent baseband nodes are susceptible to charge leaks, which induces unnecessary losses in the corresponding baseband signals. Finally, 50% duty cycle square waves generated by $N/2$ Johnson counters are limited in frequency. The frequency of the LO can be no higher than half the frequency at which a large enough voltage swing can be obtained. State-of-the-art 45nm SOI FETs can generate LO frequencies of about 30 GHz.

The MOSFETs in Fig. 6.20 can be replaced by HBTs, which requires pull-up supply voltage as well as a biasing choke to ground. This leads to the circuit in Fig. 6.21.

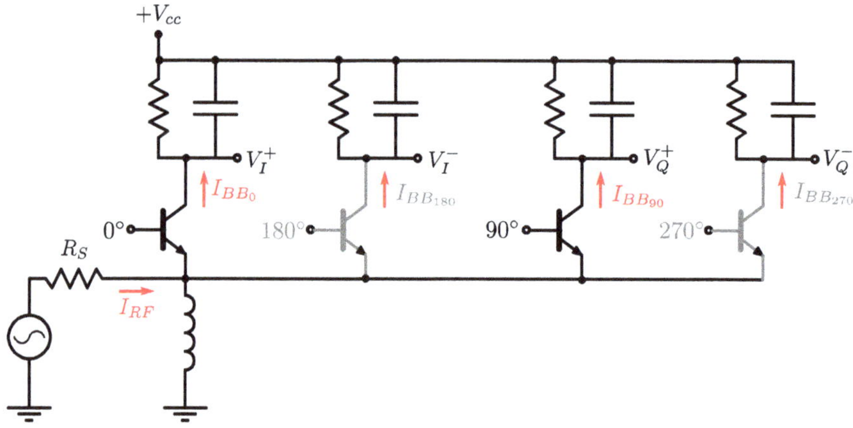

Fig. 6.21 HBT switching illustration. Devices that are conducting are shown in black, and non-conducting devices are in grey. Adapted from [25]

References

1. Shirazi AHM et al (2016) On the design of mm-wave self-mixing-VCO architecture for high tuning-range and low phase noise. IEEE J Solid-State Circuits 51(5):1210–1222. https://doi.org/10.1109/JSSC.2015.2511158
2. Razavi B (2012) RF microelectronics, 2nd edn. Pearson Education, Inc.d, Los Angeles, California
3. Spasaro M, Zito D (2019) Millimeter-wave integrated silicon devices: active versus passive-the eternal struggle between good and evil: (Invited Paper). In: Proceedings of the International Semiconductor Conference, CAS, vol 2019, no 829005, pp 11–20. https://doi.org/10.1109/SMICND.2019.8923669
4. Doan CH, Emami S, Niknejad AM, Brodersen RW (2005) Millimeter-wave CMOS design. IEEE J Solid-State Circuits 40(1):144–154. https://doi.org/10.1109/JSSC.2004.837251
5. Shahramian S et al (2011) Design of a dual W- and D-band PLL. IEEE J Solid-State Circuits 46(5):1011–1022. https://doi.org/10.1109/JSSC.2011.2117050
6. Imani A, Hashemi H (2018) Frequency and power scaling in mm-wave Colpitts oscillators. IEEE J Solid-State Circuits 53(5):1338–1347. https://doi.org/10.1109/JSSC.2017.2788861
7. Wang C, Chen Z, Heydari P (2012) W-band silicon-based frequency synthesizers. IEEE Trans Microw Theory Tech 60(5):1307–1320
8. Mangraviti G, Khalaf K, Arvais B, Vaesen K, Szortyka V (2015) Design and tuning of coupled-LC mm-wave subharmonically injection-locked oscillators. IEEE Trans Microw Theory Tech 63(7):2301–2312
9. Kamogawa K, Tokumitsu T, Aikawa M (1997) Injection-locked oscillator chain: a possible solution to millimeter-wave MMIC synthesizers. IEEE Trans Microw Theory Tech 45(9):1578–1584. https://doi.org/10.1109/22.622925
10. Liao D, Zhang Y, Dai FF, Chen Z, Wang Y (2020) An mm-wave synthesizer with Robust locking reference-sampling PLL and wide-range injection-locked VCO. IEEE J Solid-State Circuits 55(3):536–546. https://doi.org/10.1109/JSSC.2019.2959513
11. Buckwalter JF, Babakhani A, Komijani A, Hajimiri A (2006) An integrated subharmonic coupled-oscillator scheme for a 60-GHz phased-array transmitter. IEEE Trans Microw Theory Tech 54(12):4271–4279. https://doi.org/10.1109/TMTT.2006.885581

12. Natarajan A, Komijani A, Guan X, Babakhani A, Hajimiri A (2006) A 77-GHz phased-array transceiver with on-chip antennas in silicon: transmitter and local LO-path phase shifting. IEEE J Solid-State Circuits 41(12):2807–2818. https://doi.org/10.1109/JSSC.2006.884817
13. Agiwal M, Roy A, Saxena N (2016) Next generation 5G wireless networks: a comprehensive survey. IEEE Commun Surv Tutor 18(3):1617–1655. https://doi.org/10.1109/COMST.2016.2532458
14. Taori R, Sridharan A (2014) In-band, point to multi-point, mm-Wave backhaul for 5G networks. In: IEEE International Conference on Communications Workshops, pp 96–101. https://doi.org/10.1109/ICCW.2014.6881179
15. Shafi M et al (2017) 5G: a tutorial overview of standards, trials, challenges, deployment, and practice. IEEE J Sel Areas Commun 35(6):1201–1221. https://doi.org/10.1109/JSAC.2017.2692307
16. Uwaechia AN, Mahyuddin NM (2020) A comprehensive survey on millimeter wave communications for fifth-generation wireless networks: feasibility and challenges. IEEE Access 8:62367–62414. https://doi.org/10.1109/ACCESS.2020.2984204
17. Hoydis J, Ten Brink S, Debbah M (2013) Massive MIMO in the UL/DL of cellular networks: how many antennas do we need? IEEE J Sel Areas Commun 31(2):160–171. https://doi.org/10.1109/JSAC.2013.130205
18. Bao-Yen Tsui J (2005) Microwave receivers with electronic Warfare applications. Scitech Publishing, Inc., Raleigh, North Carolina
19. Ahmed A, Huang MY, Munzer D, Wang H (2021) A 43–97-GHz mixer-first front-end with quadrature input matching and on-chip image rejection. IEEE J Solid-State Circuits 56(3):705–714. https://doi.org/10.1109/JSSC.2020.3045046
20. Larsson EG, Edfors O, Tufvesson F, Marzetta TL (2014) Massive MIMO for next generation wireless systems. IEEE Commun Mag 52(2):186–195. https://doi.org/10.1109/MCOM.2014.6736761
21. Gustavsson U et al (2014) On the impact of hardware impairments on massive MIMO. In IEEE Globecom Workshop—Massive MIMO: From Theory to Practice, pp 294–300. https://doi.org/10.1109/GLOCOMW.2014.7063447
22. Song P, Hashemi H (2020) Mm-wave mixer-first receiver with passive elliptic low-pass filter. Digest of Papers—IEEE Radio Frequency Integrated Circuits Symposium, vol 2020, pp 271–274. https://doi.org/10.1109/RFIC49505.2020.9218288
23. Krishnamurthy S, Iotti L, Niknejad AM (2021) Design of high-linearity mixer-first receivers for mm-wave digital MIMO arrays. IEEE J Solid-State Circuits 56(11):3375–3387. https://doi.org/10.1109/JSSC.2021.3101984
24. Kashani MH, Tarkeshdouz A, Afshari E, Mirabbasi S (2019) A 53–67 GHz low-noise mixer-first receiver front-end in 65-nm CMOS. IEEE Trans Circuits Syst I Regul Pap 66(6):2051–2063. https://doi.org/10.1109/TCSI.2019.2895893
25. Ying R, Molnar A (2021) Impedance transparency and performance metrics of HBT-based N-Path mixers for mmWave applications. IEEE Trans Circuits Syst I Regul Pap 68(5):2210–2223. https://doi.org/10.1109/TCSI.2021.3060644
26. Razavi B (2014) Fundamentals of microelectronics, 2nd edn. John Wiley & Sons Inc., Hoboken, New Jersey

Chapter 7
High-Performance Si Data Converters for Millimeter-Wave Transceivers

Wideband communication systems rely on conversion circuits to translate between the analogue and digital domains. The analogue-to-digital (A/D) converter plays a significant role in the achievable data rate that the receiver can process. Conversely, the digital-to-analogue (D/A) converter influences the instantaneous TX bandwidth and, subsequently, the maximum data rate. A/D converter design is centred around the power-bandwidth product, which essentially requires careful balancing of the sampling rate and resolution. The design goal is always to maximise bandwidth while keeping the power consumption to a minimum. An estimated 5 resolution bits and 1 GHz resolution bandwidth is required for 60 GHz transceivers [1]. However, applications such as wireless video streaming defined in the IEEE 802.15.3c specification may require up to 8 resolution bits and sampling rates that exceed 2.5 Gs/s. High sampling rates are possible in both SiGe and CMOS technologies, and rates exceeding 30 Gs/s have been reported [2, 3], but this comes at the cost of excessive power consumption. Another issue is that the dynamic range depends on clock jitter, which is where good integration between the analogue and digital MMIC sections is extremely advantageous.

High-bandwidth D/A converters with good linearity are required in multiple mm-wave transmitter applications [4, 5]. High-speed operation favours BiCMOS processes due to their excellent frequency performance, but CMOS offers some unique advantages in this regard as well.

7.1 A/D Converters

Advances in the digital domain, especially CMOS technology, have led to many signal processing operations gradually shifting to digital form. Generally speaking, transistor scaling improves digital performance but degrades analogue performance. With that said, further scaling is not a definite solution to improve the performance

J. du Preez and S. Sinha, *State-of-the-Art of Millimeter-Wave Silicon Technology*, Lecture Notes in Electrical Engineering 945,
https://doi.org/10.1007/978-3-031-14655-8_7

of future digital systems, as power and area efficiency are important considerations. Efficiency improvement can be partially obtained by using a newer technology, but new architectures and circuit techniques are needed to push the boundaries of performance further. A/D converter researchers, therefore, have to make a fundamental decision: whether to improve algorithms that rely on digital circuits to offset the poor performance of scaled analogue circuits, or to chase analogue techniques and architectures that improve energy and area efficiency to cope with technology scaling.

ADCs are ubiquitous, as any system that digitally processes a continuous-time signal requires one. Figure 7.1 shows a 4-bit flash ADC used in a 25 Gbaud/s baseband receiver chain.

The receiver chain in Fig. 7.1 combines a 12.5 GHz low-pass programmable gain amplifier (PGA) with a 4-bit flash ADC. The converter combines scramblers and pseudo-random bit sequence (PRBS) generators to synchronise the data link to the FPGA, upon start-up. Furthermore, the receiver output is connected to a high-performance FPGA that provides multi-Gb/s serial receivers.

A variety of architectures are known, and these can be evaluated based on their resolution, sampling rate, bandwidth, power consumption, latency and chip area [7]. Regardless of the architecture, all A/D converters rely on comparing the input signal to a stable reference. Flash ADCs that use one comparator per bit aremuch faster than, for instance, a successive approximation register (SAR) ADC, which uses one comparator sequentially. However, the flash architecture becomes extremely large and inefficient in terms of area when higher resolution bits are required. To combat some of the widely known limitations of flash converters, alternatives such as interpolating and pipelined architectures have also been used.

The trade-offs between converter performance specifications favour one particular architecture for any application. For example, faster conversion is obtained by lowering the number of resolution bits. Due to its component matching requirement, Flash ADCs can realistically provide about 8 or 9 resolution bits. SAR ADCs can provide resolutions in the 8 to 16 bits range while consuming a reasonably small chip area and low power consumption. Integrating ADCs can provide the highest number of bits but can only achieve low conversion speeds, limiting their use to applications such as portable instruments where only low-bandwidth signals are measured.

7.1.1 Architectures

7.1.1.1 Flash

The simplest method of N-bit A/D conversion is to compare a sample-and-hold (S/H) version of an analogue input to 2^N reference voltages [7–9]. The flash ADC, which consists of resistive dividers along with a binary encoder and a comparator array, is derived from this principle. The S/H circuit holds the analogue input for half of the clock cycle. Figure 7.2 shows a flash ADC circuit.

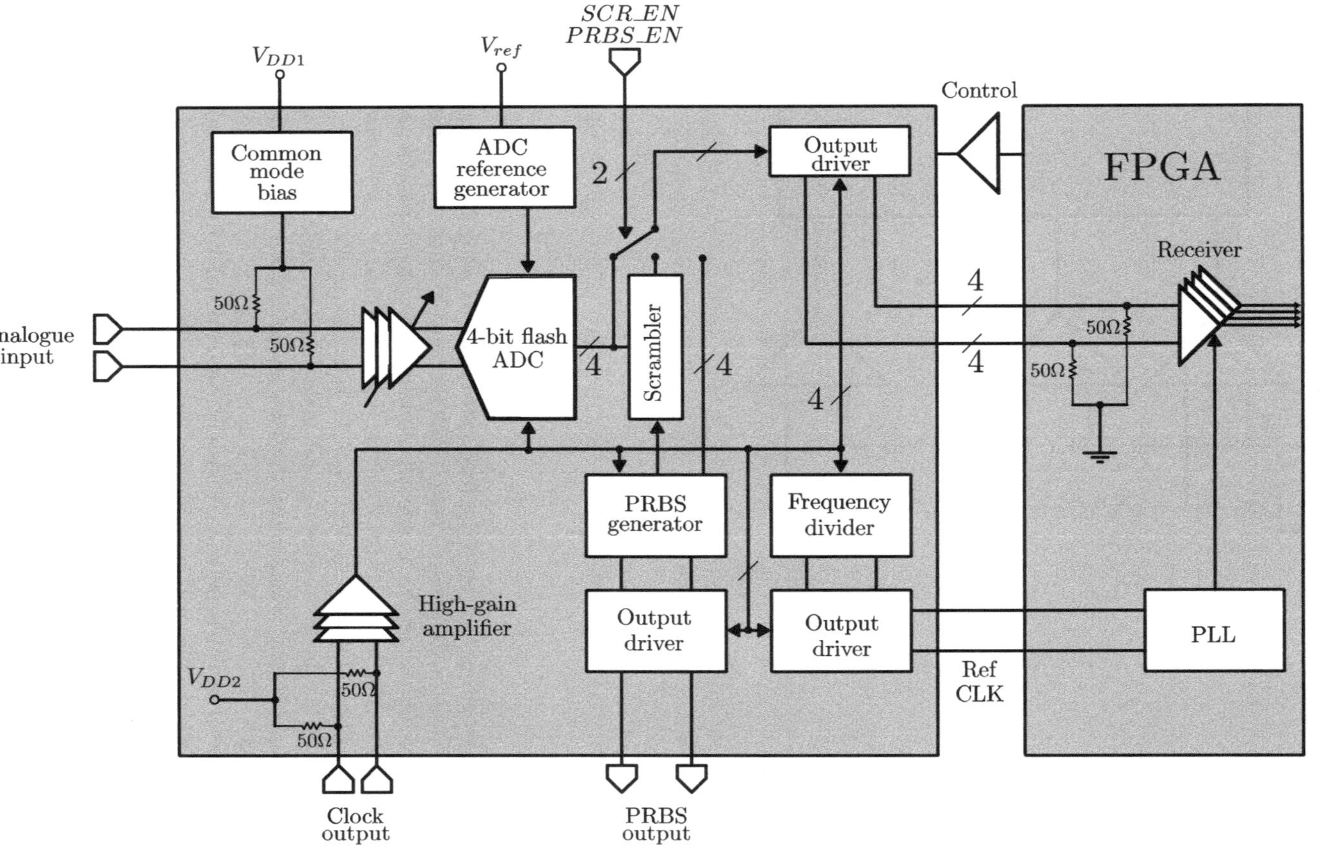

Fig. 7.1 25 Gbaud/s baseband receiver chain, adapted from [6]

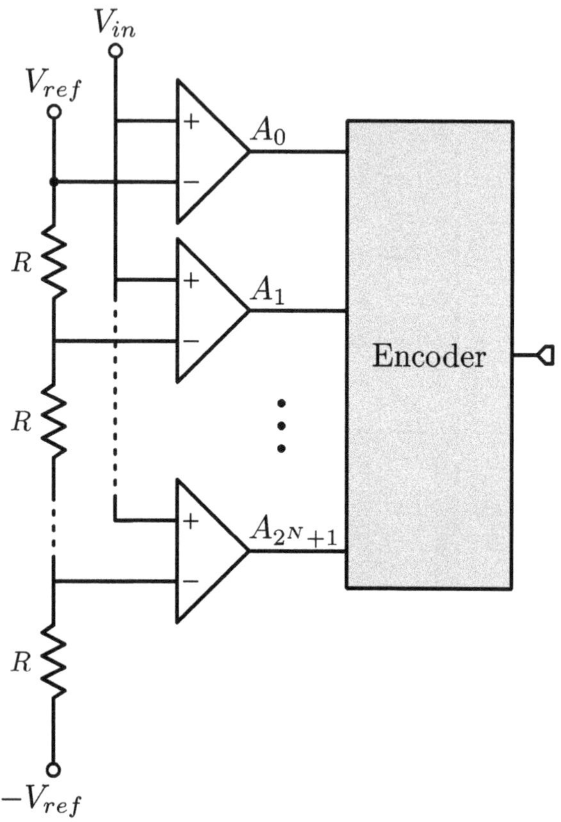

Fig. 7.2 Flash ADC circuit

As seen from Fig. 7.2, an N-bit flash converter uses $2^N + 1$ comparators. Each comparator uses a different reference voltage derived from the resistor divider chain. The input voltage V_{in} is applied to the non-inverting input terminal of each comparator, and their outputs are Boolean. Each bit (A_0, A_1, ...) indicates whether the input voltage is greater or less than the reference at the relevant comparator. The output of this circuit is sometimes called thermometer code, since connecting its outputs to an LED array would mimic the rising and falling action of an old-fashioned thermometer.

The number of comparators in a flash converter doubles for every added bit of resolution, which is the main reason why flash converters in general, and especially mm-wave ones, are limited to less than $4 - 6$ bits. Mismatched comparators that result from differing propagation delays can cause incorrect readings. One method to combat this is to use an S/H circuit that precedes the system, but this would degrade the speed for which the flash architecture is known and used.

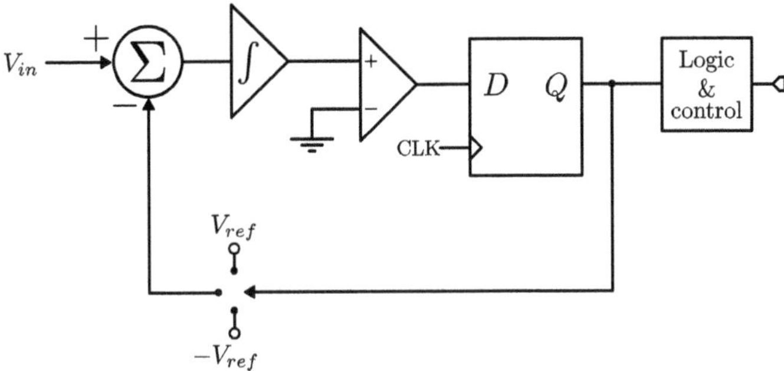

Fig. 7.3 Delta-sigma converter

7.1.1.2 Delta-Sigma

The naming of the delta-sigma or $\Sigma - \Delta$ converter implies that it operates on sum and difference voltages. This is basically true, even though the converter operation is more complex than the name suggests. The delta-sigma converter uses a feedback loop that tracks changes in the input signal, similar to the SAR ADC [7, 9]. Figure 7.3 shows a block diagram of a simplified delta-sigma converter.

This type of converter consists of a delta-sigma modulator followed by a digital logic subsystem. The modulator converts the analogue input into a digital waveform, and the digital subsystem generates a usable digital output signal. The input signal drives one port of a summing junction, the output of which goes through an integrator into a comparator. The inverting input of this comparator is grounded, meaning that its output will simply indicate whether the integrated signal at the non-inverting input is positive or negative.

A D-type flip flop samples the digital output of the comparator. The flip flop will output the value at its D input to its Q output on a rising edge. The clock input frequency corresponds to the system's highest possible sampling rate, and the Q output is split between the digital logic system and an analogue switch. The switch toggles between the positive and negative reference voltages and drives the negative feedback loop through the difference port of the summing junction. Moreover, the negative feedback effect means that the system continuously tries to cancel out integrated effects of the input by generating pulses with the opposite polarity via the switch. In some cases, the input to the summing junction is first passed through an S/H circuit.

7.1.1.3 Successive Approximation Register

A SAR ADC is shown in Fig. 7.4. The circuit consists of a clock generator, comparator, digital control logic, SAR, DAC and a latched output buffer [7].

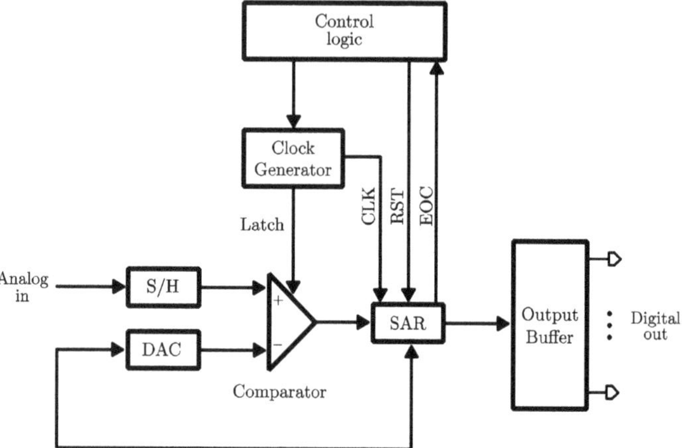

Fig. 7.4 Successive approximation ADC block diagram

The digital control logic handles the bit cycling, clock enable and reset functions. The SAR ADC operates based on a binary search mechanism to find the closest digital output code corresponding to the input voltage. This principle uses various DAC input codes to compare to the analogue input, starting with the most-significant bit (MSB) where each bit is set to a logic 1. If the analogue input is greater than the voltage generated by the DAC (which is half the full-scale value), the starting value of the MSB is retained and the process continues with the next DAC output (which is $3/4$ of the full-scale value). If not, the MSB is flipped to a logic 0 and in this case, the DAC output for the next step will be $1/4$ of the full-scale value. This process repeats until the LSB is reached, which effectively means that the DAC output is within $\pm 1/2$ LSB of the analogue input voltage. An N-bit SAR ADC has 2^N resolution levels, corresponding to a minimum of N clock periods required for the conversion.

7.1.2 Mm-Wave A/D Converters

7.1.2.1 High-Performance Mm-wave Considerations

High-performance digital communication systems require a higher effective number of bits (ENOB) and sampling rate performance than ever before. Upping the converter sampling rate typically leads to pipelining multiple stages to relax the analogue devices' requirements. In this case, a low-resolution quantiser is followed by a low-resolution multiplying DAC.

Next-generation fibre-optic systems will operate at bandwidths exceeding 100 Gb/s, requiring A/D and D/A converters with 32 GHz bandwidth or more [4].

Such a high converter bandwidth is challenging for even the most advanced FinFET technologies. Generally, good SNDR can be achieved in highly scaled technologies, but the high bandwidth remains a difficult challenge to overcome. Highly scaled 7 nm and 14 nm FinFET technologies are great options for parallel SAR architectures because of their small footprint and low power consumption [10]. However, these devices suffer from large parasitic capacitances in the metal layers, resulting in poor f_{max} and limiting converter bandwidth. As such, scaling down from 65 nm bulk CMOS all the way down to 14 nm FinFET has seen little progress made in terms of signal processing bandwidth. On the other hand, SiGe HBT performance has shown better scaling potential.

7.1.2.2 Flash

The high-speed capability of flash converters makes them excellent candidates for mm-wave systems. A number of useful architectures have been reported for flash ADCs in mm-wave systems [3, 11]. The simplest mm-wave technique used is the time-interleaved architecture, as shown in Fig. 7.5.

The time-interleaved architecture relies on parallel ADC cores. As such, they must be calibrated periodically for skew, gain mismatch and offset. CMOS implementations often use time-interleaved THAs that drive multiple banks of time-interleaved sub-ADCs. SiGe amplifiers are also used since they can drive high capacitive loads.

Speed improvements can be made by driving the comparator bank as if it were a lumped load, as shown in Fig. 7.6 [12].

A 40 GHz sampling rate was achieved by Cheng et al. with this architecture, using a SiGe BiCMOS technology with $f_T/f_{max} = 210/310$ GHz [12]. Schvan et al. used

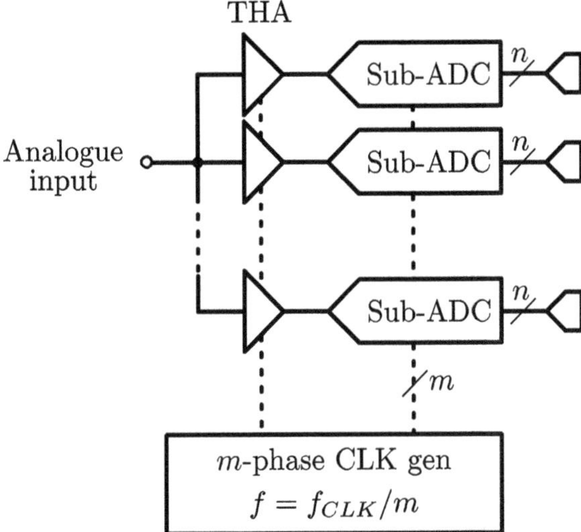

Fig. 7.5 Time-interleaved flash ADC

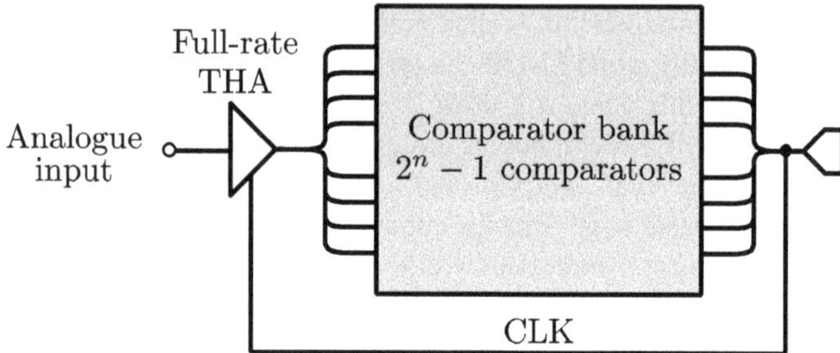

Fig. 7.6 Direct comparator driving flash ADC

a long transmission line to absorb the input capacitance of the comparator bank in their design (Fig. 7.7) [13].

This implementation does not need a THA, but the data and clock paths must be carefully matched to avoid skew. A variant of this architecture is reported by Shahramian et al. [3]. A block diagram of this implementation is shown in Fig. 7.8.

This design consists of a tree of buffer amplifiers that drives the capacitive load. A transimpedance amplifier is placed on the front-end and used as a low-noise gain block. Coupled with the buffer amplifier tree, this ADC can process very small amplitudes (0.24 V_{pp} differential). The data and clock trees are symmetric to minimise skew. The ADC achieves 35 GS/s sampling rate with 4 resolution bits, $SFDR$ of

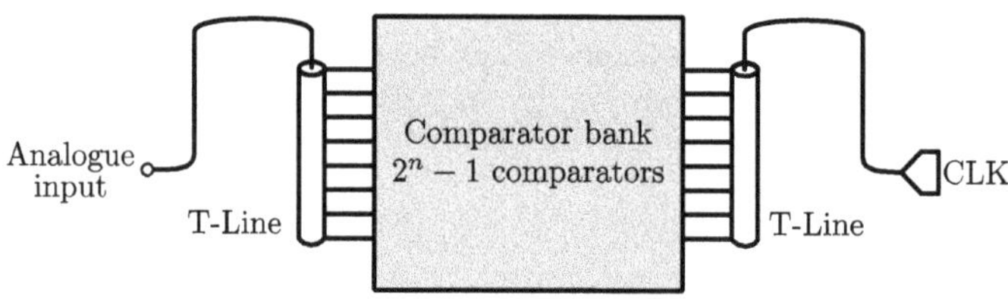

Fig. 7.7 Transmission line distribution of clock and data in a flash ADC

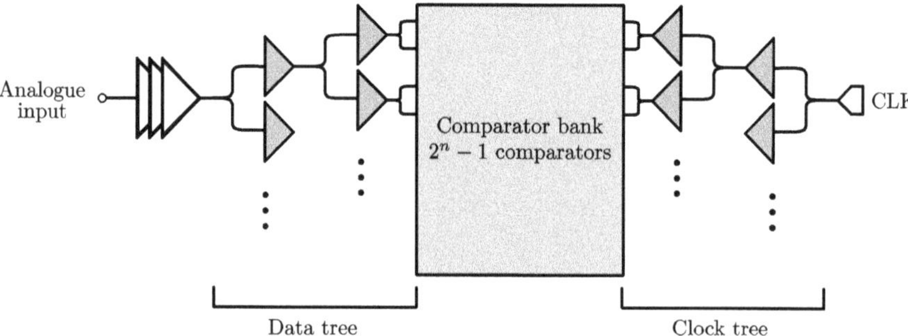

Fig. 7.8 Active distribution network flash ADC

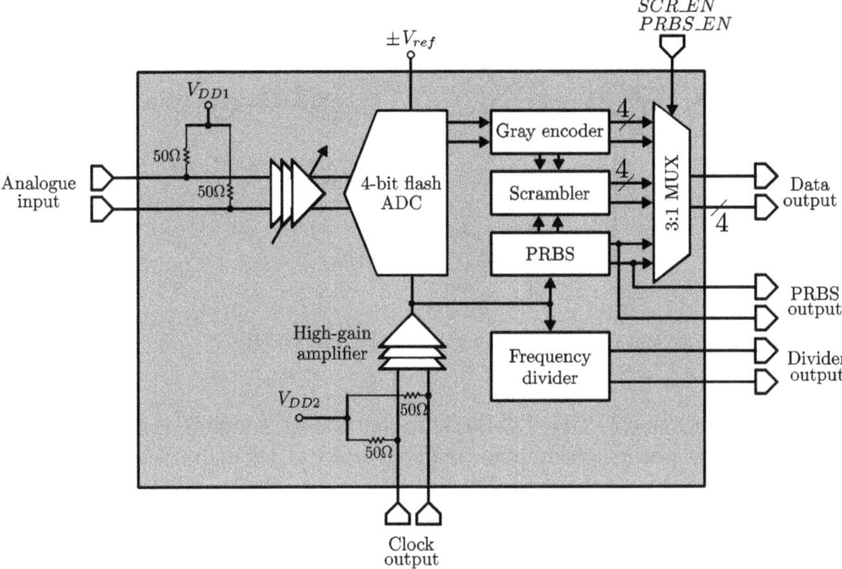

Fig. 7.9 Mm-wave flash ADC architecture

27.3 dBc at 11 GHz, 4.5 W 0.18 μm SiGe BiCMOS process (SBC18HX from Jazz Semiconductor). This process offers modest f_T 160 GHz f_T 50 GHz.

A mm-wave flash ADC proposed by Du et al. is shown in Fig. 7.9 [11]. This converter is implemented in a standard 0.13 μm SiGe BiCMOS technology.

This converter consists of a 4-bit flash ADC core coupled with a Gray encoder to suppress bubble errors. The flash core uses a travelling-wave topology, which circumvents the need for a THA in the front-end chain. In this arrangement, the input clock and analogue signals propagate synchronously between comparators. The outputs of the ADC are Gray encoded to facilitate further DSP down the line. The combination of the 3 : 1 multiplexer, scrambler and PRBS generator allows ADC sample storage on an FPGA. The flash ADC core and the Gray encoder determine the converter linearity. Finally, this converter achieves $ENOB \geq 3$ and $SFDR \geq$ 24.8 dBc at a peak sampling rate of 39.04 GS/s and DC to 20 GHz bandwidth.

7.1.2.3 Delta-Sigma

Delta-sigma ADCs have found application in mm-wave systems. An example receiver slice from a 16-element beamformer reported by Lu et al. is shown in Fig. 7.10 [14].

Each slice of the front end contains an LNA, 27 GHz LO with buffer, passive mixer, VGA, an ADC array driven by a delay-locked loop (DLL) and finally, a bitstream adder. The design uses a 1 GHz IF to facilitate digital I/Q mixing, whereas the RF

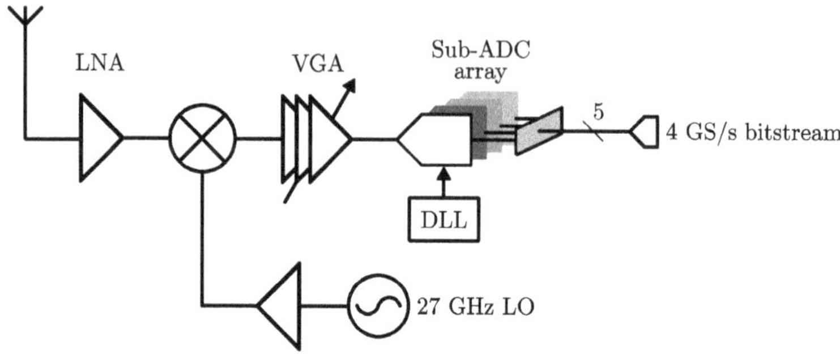

Fig. 7.10 Single slice in a beamforming receiver, adapted from [14]

mixer only needs one 27 GHz LO for downconversion. A high IF is desirable, since it leads to lower power consumption and chip area use due to the ease of generating and distributing the single-phase LO. Moreover, it facilitates AC coupling so that common-mode blocks can be independently optimised. The summed outputs of the ADC array feed a digital beamformer at 4 GS/s.

The parallel ADC array is shown in greater detail in Fig. 7.11.

The converters used are continuous-time bandpass delta-sigma ADCs. These ADCs are resilient to aliasing, easing the implementation of the input filter. They also provide resistive inputs, which are easier to drive. Such a modulator is shown in Fig. 7.12.

The block diagram in Fig. 7.12 basically describes a cascaded resonator-feedback bandpass sigma-delta modulator with an added feedforward path. Each modulator uses two single-opamp resonators that are both area and power efficient. One half of the feedback DACs are return-to-zero (RZ) and the other half are half-delay RZ pulse-shaped current-steering DACs. The 4 GHz master clock is distributed

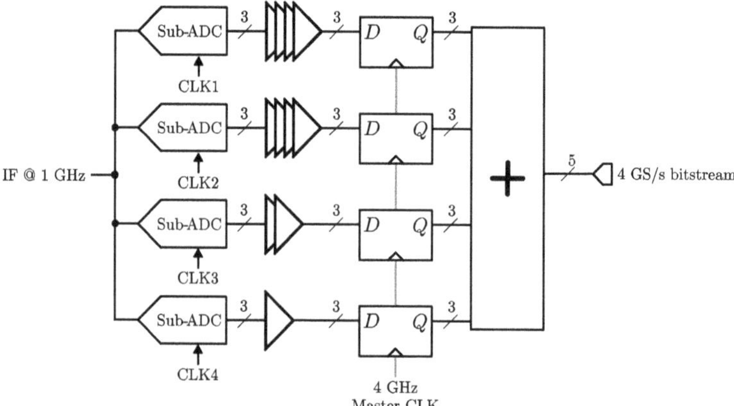

Fig. 7.11 Parallel ADC architecture

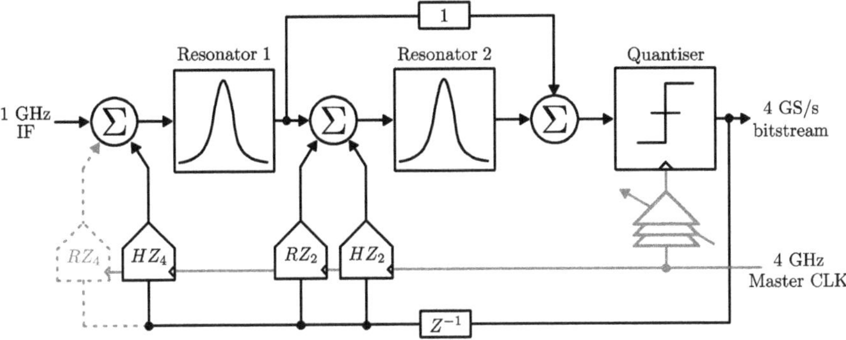

Fig. 7.12 Continuous-time bandpass sigma-delta modulator

to converters and beamformer logic alike, and each of the ADCs generates a 3-bit, 4 GS/s output stream. These are aligned (digitally) and summed, resulting in the 5-bit, 4 GS/s output. The whole chip is implemented in a 40 nm CMOS technology, since the remainder of the beamformer contains significant digital logic and processing subsystems. Moreover, CMOS technology scales very well with oversampling ADCs, and more advanced technology nodes can provide even higher sampling rates and bandwidths.

7.1.2.4 Successive Approximation Register

Figure 7.13 shows a time-interleaved SAR ADC implementation proposed by Zandieh et al. [15]. The ADC is implemented in a 22 nm FD-SOI CMOS node.

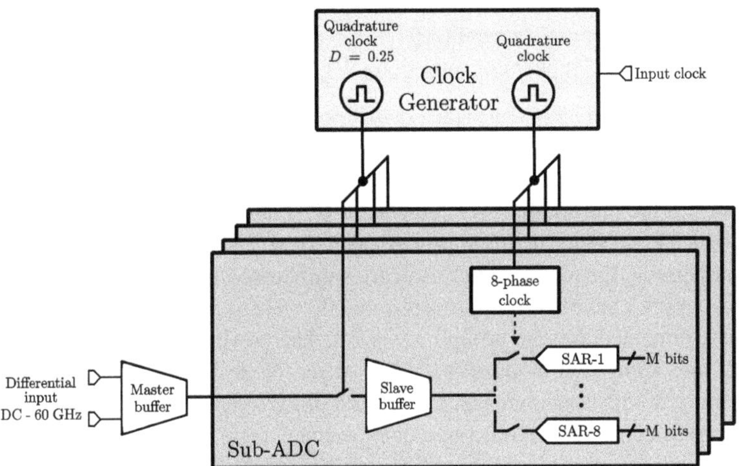

Fig. 7.13 High-speed mm-wave SAR ADC, adapted from [15]

The ADC resolution can vary from 6 to 8 bits and achieves 128 GS/s speed. A master buffer drives four track-and-hold (T/H) circuits operated with a 25% duty cycle. The clock is provided by a quadrature clock generator, which provides the 25% duty cycle clock at 32 GHz. Using staggered clocks can reduce the load at the output of the master buffer, thereby increasing the overall front-end bandwidth. The master T/H circuits are all followed by eight slave T/H circuits driven by 12.5% duty cycle, 8-phase clock sources at 4 GHz. These are synchronised to the master clock using another quadrature source that generates 32 GHz, 50% duty cycle clocks. The 32 M-bit sub-ADCs are replaced with an output termination for S-parameter measurement and testing with large-signal waveforms.

7.1.2.5 Auxiliary Amplifiers

The transimpedance amplifier (TIA) is a central building block of broadband ADCs. In a typical modern fibreoptic system, the TIA is tasked with amplifying complex-modulated signals. As such, they need to be wideband, low noise and linear with low power consumption and high gain, quite a substantial list of requirements. The linearity of the TIA is described through its total harmonic distortion (THD) as well as its linear input voltage or current range. The TIA FoM is defined as [4]

$$FoM_{TIA} = \frac{Z \cdot I_{max} \cdot BW}{i_n^{rms} \cdot P \cdot THD} \tag{7.1}$$

where:

- Z is the transimpedance gain (Ω)
- I_{max} is the maximum linear input current (A)
- BW is the 3-dB bandwidth (Hz)
- i_n^{rms} is the input noise current (A)
- P is the total DC power consumption (W)

To compare the TIA performance in different technology nodes, Voinigescu et al. designed a benchmark circuit, shown in Fig. 7.14 [4].

This amplifier is designed to handle a 300 mV$_{pp}$ input signal, and both devices are biased at the minimum noise figure at 90 GHz and (±1 mA/μm). Moreover, it is based on a 55 nm SiGe BiCMOS process with $f_T = 280$ GHz and $f_{max} = 430$ GHz. For comparison, the technologies used are summarised in Table 7.1.

The amplifier bandwidth improved from 100 GHz to 250 GHz across nodes while the noise remained similar, around $3 - 4$ dB. The power consumption in Node 1 is 26 mW and 40 mW in Nodes 2 to 5. Moreover, Node 5 with 250 GHz bandwidth can achieve 10 dB transimpedance gain with noise figure below 5 dB.

Voinigescu et al. carried out a similar experiment for a T/H amplifier (THA). THAs offer a solution to the requirement for very high bandwidth samplers used in next-generation ADCs. CMOS switches can achieve 100 GHz bandwidth with relative

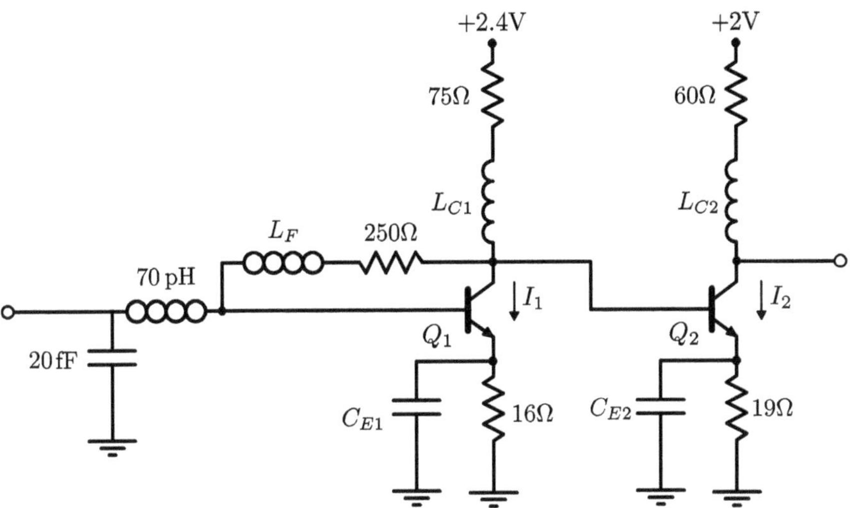

Fig. 7.14 Benchmark transimpedance amplifier circuit

Table 7.1 High-frequency performance scaling of NMOS and SiGe HBT devices

Parameter	Node 1	Node 2	Node 3	Node 4	Node 5
FET L_G (nm)	20	14	10	7	5
HBT W_E (nm)	115	88	66	44	22
NMOS f_T (GHz)	334	436	609	423	210
HBT f_T (GHz)	282	395	502	642	778
HBT f_{max} (GHz)	432	543	766	1191	1985

ease, but they rely on rail-to-rail switching and other CMOS logic systems that cannot operate with such high clock frequencies. In this regard, bipolar THAs have excellent scaling with frequency. A THA FoM should combine tracking bandwidth, spurious-free dynamic range (SFDR), the system clock frequency, gain, linearity, power consumption and compression point. A suitable FoM is

$$FoM_{THA} = \frac{G \cdot IP_1 \cdot BW \cdot SFDR \cdot f_{CLK}}{P} \tag{7.2}$$

The experimental Node 1 implementation achieved just over 70 GHz bandwidth with f_{CLK} exceeding 100 GHz. In Node 5, the bandwidth improved to 92 GHz, and it can be further increased by either doubling the current or halving the holding capacitor value. Retaining the same holding capacitance is ideal because in this case the SFDR remains consistent and the improved bandwidth is up to 140 GHz. Doubling the current doubles the power consumption from 42 mW to 84 mW.

7.2 D/A Converters

D/A converters facilitate the interface between digital systems and the real world. They transform a binary input code into an analogue output signal, which can be either voltage or current. Ideally, waveforms ranging from DC to the Nyquist frequency ($f_s/2$) can be generated but practically, converter performance tends to deteriorate rapidly when approaching $f_s/2$. Nyquist DACs can be implemented in a number of architectures—binary-weighted, segmented and thermometer-coded—and use multiple different conversion techniques. Resolution of high-speed DACs is limited primarily by circuit complexity and chip area. Segmented DACs attempt to overcome this by using an initial stage to decode the MSB and a secondary stage to decode the remaining LSBs. Linearity and conversion monotonicity are additional limiting characteristics that complicate the design of high-resolution DACs. As such, most integrated DACs are designed with built-in calibration circuits to overcome process variations.

7.2.1 Operating Principles

The D/A converter in Fig. 7.15 combines the voltage reference (V_{REF}) with an n-bit input word to produce an analogue output [8].

The full-scale voltage V_{FS} (or current, I_{FS}) is dependent on the voltage reference and converter gain, which is often set to unity. Additionally, the voltage and current offset (V_{OS} and I_{OS}, respectively), denote the output voltage or current when a zero input word is present.

DAC characteristics to take note of are speed, resolution, accuracy and monotonicity. The speed of the DAC corresponds to the number of voltage samples it can convert per second. Most DACs need multiple clock cycles to perform a conversion, and the maximum speed is thus much less than the maximum clock frequency. Complications such as oversampling and decimation can arise because of this. Accuracy is defined as the error between the intended analogue output (given a binary input code) and the measured output. It can be specified as an absolute percentage, as is the

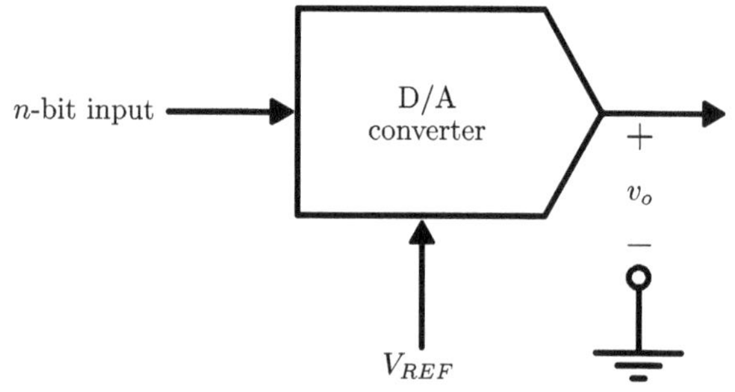

Fig. 7.15 Voltage-input D/A converter

case when the output is compared to a known voltage standard (such as transistor-transistor logic (TTL) or high-speed transceiver logic (HSTL) used in FPGA memory interfaces). The calibration for standardised interfaces can be easily obtained from libraries. Relative accuracy is specified in terms of non-linearity. It can be measured in multiple ways, and the most common method of characterisation is to note changes relative to a perfect straight line. Alternatively, the amplitude of spurious frequency domain components when the DAC should output a single tone can be used to describe non-linearity.

Monotonicity refers to a characteristic where the DAC transfer function only ever increases, and every DAC is required to exhibit this. Modern integrated DACs almost universally guarantee monotonicity. DAC resolution is the same as for an ADC and indicates the voltage change per LSB.

7.2.2 *Architectures*

Transforming an N-bit digital code into an analogue output relies on a series of switches to select the correct reference voltage. These voltages are then summed to obtain a discrete-time output signal. This output, in turn, is low-pass filtered for smoothing. The technology and switching scheme heavily influence the converter performance [9].

7.2.2.1 R-2R Ladder

The R-2R ladder DAC is used in some exotic designs, but their utilisation has diminished considerably in favour of newer CMOS-based designs. Nonetheless, it provides an intuitive baseline from which to approach further DAC circuits. One such R-2R ladder is shown in Fig. 7.16.

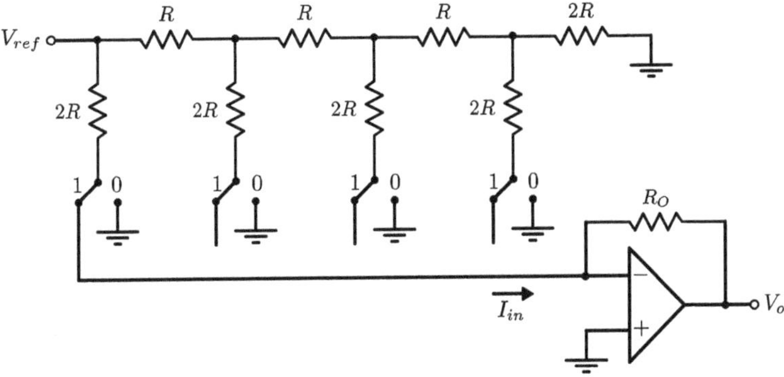

Fig. 7.16 R-2R ladder DAC

The circuit utilises a periodic ladder, which divides the input voltage to each stage by 2, depending on the accuracy of the resistors. The output voltage is

$$V_o = -\frac{V_{ref} R_o}{2R}\left(\frac{A_3}{2^0} + \frac{A_2}{2^1} + \frac{A_1}{2^2} + \frac{A_0}{2^3}\right) \tag{7.3}$$

where A_n represents bits of the digital input code. More bits can of course be added, provided that the resistor accuracy can be maintained to ensure a divide-by-2 ratio.

7.2.2.2 Binary-Weighted

The easiest and most area-efficient switch control method is direct actuation via digital I/O. However, the components determining the reference voltage levels must be binary weighted, which is perhaps the main drawback of this method. For high-precision DACs, it means that the components must be matched to within $\pm 1/2$ LSB. The output signal can often exhibit a large spike when large code transitions are present, as is the case when moving from 0111 to 1000. This happens as a result of asymmetrical switching, where the LSBs can all switch at a higher rate than the MSB. The DAC output will accordingly try to return to zero first, before assuming the correct state, which is effectively a glitch. The energy of such glitches can be measured to characterise their effect, and this is commonly expressed as pV/s for high-precision DACs.

7.2.2.3 Thermometer-Coded

A thermometer-coded DAC uses 2^N elements connected to switches. The switches are controlled from a binary-to-thermometer decoder. When the input code changes by a single LSB, only one element needs to be switched. Glitch energy is substantially less, using this method. Moreover, since the output follows the changes in the input code, monotonicity is guaranteed.

7.2.2.4 Switched-Capacitor

The switched-capacitor DAC improves the R-2R ladder and is much better suited for integrated designs. Figure 7.17 shows a circuit.

The circuit operates on two opposing phases, indicated by ϕ_1 and ϕ_2. During the first phase, all of the ϕ_1 switches are closed and the ϕ_2 switches are open. The capacitors connected between each switch pair are selected to have binary-weighted values. That is, $C_2 = 2C_1$, $C_3 = 2C_2$, up until the last capacitor $C_N = 2C_{N-1}$. Finally, the value of the feedback capacitor is $C_1(2^N - 1)$, where N equates to the number of resolution bits.

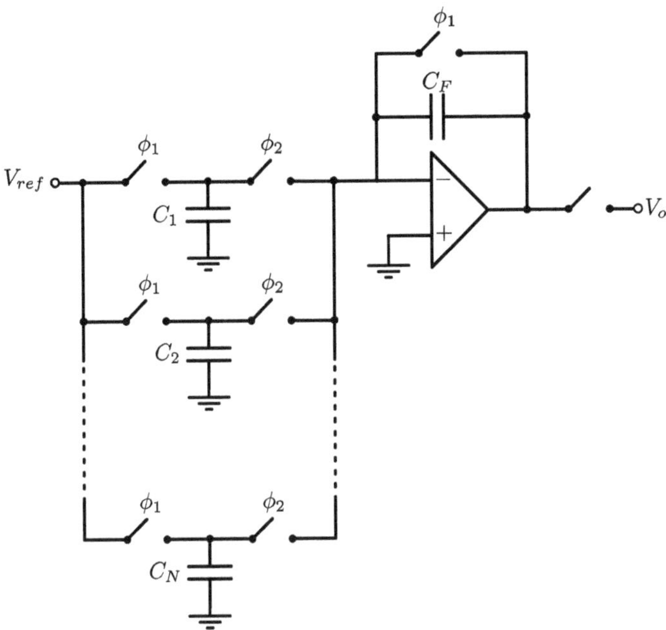

Fig. 7.17 Switched-capacitor DAC circuit

When all the ϕ_1 switches are closed, the capacitors C_1 through C_N are charged up to V_{ref}. Meanwhile, the opamp, in trying to keep its input terminals at the same potential, will charge C_F to the same total charge as is dumped into its inverting input. The speed of the opamp and the capacitor sizes are the main limitations to conversion speed. As soon as the opamp output is settled, the output switch ϕ_2 will close, allowing the analogue voltage to propagate to the output port.

7.2.2.5 Single-Bit DAC

The single-bit DAC (or one-bit DAC) is predominantly digital and consists of one switch that alternates between V_{ref} and ground. The switch is driven digitally to generate an output whose duty cycle is proportional to the desired output. The duty cycle is average over a period $\tau = RC$, derived from the passive RC filter at the switch output. This filter also keeps harmonics under control, to an extent. The switching frequency of the single-bit DAC is significantly higher than the maximum desired output frequency. An oversampling factor of 100 or more is often used. Finally, a pulse-width modulation (PWM) source often drives the switch.

7.2.3 Mm-wave D/A Converters

7.2.3.1 Challenges

CMOS scaling beyond 32 nm has provided many advantages for digital circuits. Transistors with f_T and f_{max} greater than 300 GHz have become commonplace, opening the door for a host of low-power applications in the sub-100 GHz region [16–18]. However, a major problem with scaled technologies is the reduction in breakdown voltage, which leads to lower output power and poor efficiency. As such, scaling transceivers into the 10 Gb/s and higher regime poses significant challenges, and clever circuit design techniques must be employed to reach Watt-level output power and good PAE at high data rates ($\geq$ 10Gb/s).

The high data rates required for mm-wave DACs can cause serious issues with I/O bandwidth. As an example, consider that an 8-bit, 32 GS/s DAC requires input data supplied at 256 Gb/s. Most DACs of even moderate speed solve this issue by multiplexing a parallel input stream received from an FPGA [19]. The degree of parallelisation is limited by the data transfer rate that can be provided off-chip, typically around 500 Mbps for a single low-voltage differential signalling (LVDS) channel. Meeting the aforementioned target data rate would need 512 LVDS channels. This will result in a significant increase in chip area and complexity. The I/O available on the driving FPGA is limited as well, and the LVDS channels must be synchronised. High-speed serialiser-deserialiser (SerDes) interfaces are a potential solution, but current-generation FPGAs are limited to a few Gb/s throughput. Adding multiple SerDes channels may be a possibility, but synchronising the clock between channels becomes very challenging, since these transceivers use clock and data recovery (CDR) schemes where the clock is embedded within the data stream. Multiplexing and data interleaving seem to be a solid choice in mm-wave DACs.

7.2.3.2 Benchmark Circuits

Balteanu et al. propose an array of 4 identical 2-bit DACs to form an 8-bit IQ power DAC [5]. The 2-bit DAC cell is shown in Fig. 7.18.

The input signal is split with a 180° balun to perform single-ended to differential conversion. After the balun, a differential broadband TIA amplifies the signal before phase adjustment. The phase adjustment is used to calibrate out I/Q phase imbalance of up to 6.5°, and is common practice in I/Q signal chains [20–22]. A Class D predriver consisting of a binary phase-shift keying (BPSK) modulator and a CMOS inverter follows, which drives the subsequent output stage into saturation. Amplitude mismatch between the I and Q channels is eliminated by rail-to-rail operation of the switching predriver.

Large-signal measurements have confirmed BPSK up to 5 Gb/s with a 45 GHz carrier frequency and simultaneous 2 Gb/s ASK and 2 Gb/s BPSK. Moreover, the output power in saturation is as high as 24.3 dBm with PAE peaking at 19.6% when

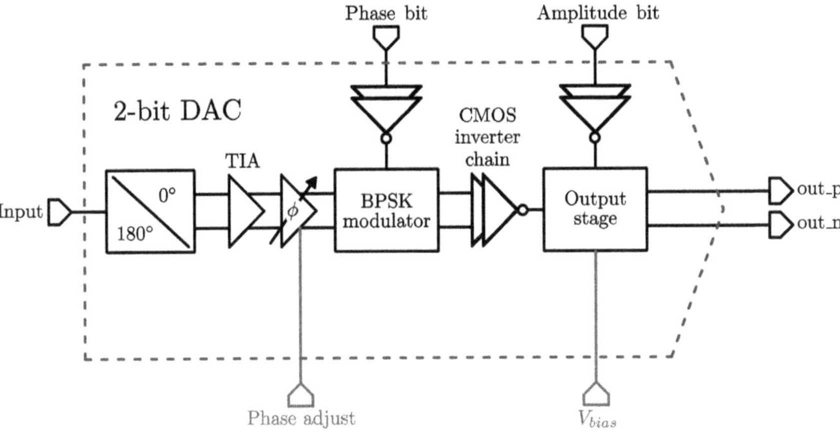

Fig. 7.18 2-bit power DAC cell

operated from a 4.4 V supply. The DAC is implemented in a 45 nm CMOS SOI technology.

Another power DAC implemented in 45 nm SOI CMOS is shown in Fig. 7.19.

The LO input is amplified and BPSK modulated prior to reaching a multi-bit modulator in the output stage. Phase modulation is done before amplitude modulation, so that the output stage is guaranteed to be driven into saturation. The two-stage implementation is advantageous because the output stage itself is simpler and provides better efficiency with lower interconnect capacitance. Additionally, the positive and negative quadrants have better symmetry, which is crucial for generating higher-order modulations accurately.

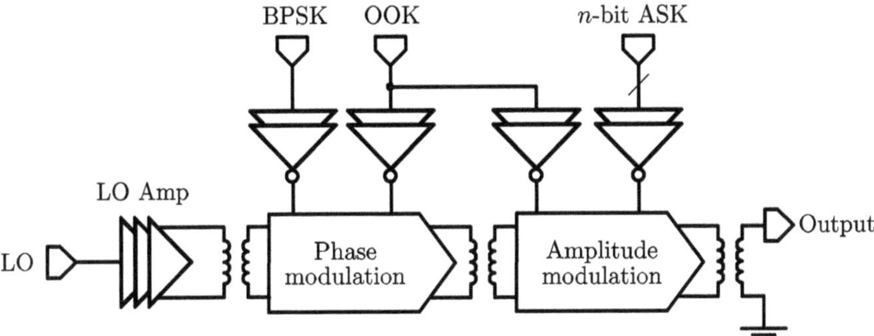

Fig. 7.19 Two-stage power DAC

7.3 Conclusion

This chapter detailed the state-of-the-art in A/D and D/A converters for mm-wave transceiver systems. The general trend throughout this chapter is that the choice between SiGe BiCMOS and CMOS is highly application dependent, with each offering unique advantages. Raw speed and bandwidth will always favour SiGe technologies, whereas complex digital processing chains and large chip integration favour the more advanced CMOS that have surfaced in recent years.

References

1. Rappaport TS, Murdock JN, Gutierrez F (2011) State of the art in 60-GHz integrated circuits and systems for wireless communications. Proc IEEE 99(8):1390–1436. https://doi.org/10.1109/JPROC.2011.2143650
2. Chu M, Jacob P, Kim JW, Leroy MR, Kraft RP, McDonald JF (2010) A 40 Gs/s time interleaved ADC using SiGe BiCMOS technology. IEEE J Solid-State Circuits 45(2):380–390. https://doi.org/10.1109/JSSC.2009.2039375
3. Shahramian S, Voinigescu SP, Carusone AC (2009) A 35-GS/s, 4-Bit flash ADC with active data and clock distribution trees. IEEE J Solid-State Circuits 44(6):1709–1720. https://doi.org/10.1109/JSSC.2009.2020657
4. Voinigescu SP, Shopov S, Bateman J, Farooq H, Hoffman J, Vasilakopoulos K (2017) Silicon millimeter-wave, terahertz, and high-speed fiber-optic device and benchmark circuit scaling through the 2030 ITRS horizon. Proc IEEE 105(6):1087–1104. https://doi.org/10.1109/JPROC.2017.2672721
5. Balteanu A et al (2013) A 2-bit, 24 dBm, millimeter-wave SOI CMOS power-DAC cell for watt-level high-efficiency, fully digital m-ary QAM transmitters. IEEE J Solid-State Circuits 48(5):1126–1137. https://doi.org/10.1109/JSSC.2013.2252752
6. Grözing M, Huang H, Du X-Q, Berroth M (2016) Data converters for 100 Gbit/s communication links and beyond. In: IEEE 16th Topical Meeting on Silicon Monolithic Integrated Circuits in RF Systems (SiRF), pp 104–106. https://doi.org/10.1109/SIRF.2016.7445481
7. Ndjountche T (2017) CMOS analog integrated circuits: high-speed and power-efficient design. CRC Press
8. Stephan K (2015) Analog and mixed-signal electronics. Wiley
9. Harpe P, BaschirottoKofi A, Makinwa AA (2014) High-performance AD and DA converters, IC design in scaled technologies, and time-domain signal processing. Springer International Publishing
10. Carter R et al (2017) 22nm FDSOI technology for emerging mobile, Internet-of-Things, and RF applications. Technical Digest—International Electron Devices Meeting, IEDM, pp 2.2.1–2.2.4. https://doi.org/10.1109/IEDM.2016.7838029
11. Du XQ, Grozing M, Buck M, Berroth M (2017) A 40 GS/s 4 bit SiGe BiCMOS flash ADC. In: Proceedings of the IEEE Bipolar/BiCMOS Circuits and Technology Meeting, pp 138–141. https://doi.org/10.1109/BCTM.2017.8112929
12. Cheng W et al (2003) A 3b 40GS/s ADC-DAC in 0.12μm SiGe. In: IEEE International Solid-State Circuits Conference, pp 374–377. https://doi.org/10.1109/isscc.2004.1332694
13. Schvan P, Pollex D, Wang SC, Fait C, Ben-Hamida N (2006) A 22GS/s 5b ADC in 0.13μm SiGe BiCMOS. In: IEEE International Solid-State Circuits Conference, pp 122–123
14. Lu R, Weston C, Weyer D, Buhler F, Lambalot D, Flynn MP (2021) A 16-element fully integrated 28-GHz digital RX beamforming receiver. IEEE J Solid-State Circuits 56(5):1374–1386. https://doi.org/10.1109/JSSC.2021.3067504

15. Zandieh A, Weiss N, Nguyen T, Haranne D, Voinigescu SP (2018) 128-GS/s ADC front-end with over 60-GHz input bandwidth in 22-nm Si/SiGe FDSOI CMOS. In: IEEE BiCMOS and Compound Semiconductor Integrated Circuits and Technology Symposium, pp 271–274. https://doi.org/10.1109/BCICTS.2018.8550842
16. Floyd B et al (2007) Silicon millimeter-wave radio circuits at 60–100 GHz (invited). In: Topical Meeting on Silicon Monolithic Integrated Circuits in RF Systems, pp 213–218. https://doi.org/10.1109/SMIC.2007.322823
17. Katayama K, Takano K, Amakawa S, Hara S, Yoshida T, Fujishima M (2016) CMOS 300-GHz 64-QAM transmitter. In: IEEE MTT-S International Microwave Symposium Digest, pp 2–5. https://doi.org/10.1109/MWSYM.2016.7540218
18. Boes F et al (2015) Multi-gigabit E-band wireless data transmission. In: IEEE MTT-S International Microwave Symposium, pp 1–4. https://doi.org/10.1109/MWSYM.2015.7166930
19. Khalil W, Wilson J, Dupaix B, Balasubramanian S, Creech GL (2012) Toward millimeter-wave DACs: challenges and opportunities. In: IEEE Compound Semiconductor Integrated Circuit Symposium (CSICS), pp 1–4
20. Marcu C et al (2009) A 90 nm CMOS low-power 60 GHz transceiver with integrated baseband circuitry. IEEE J Solid-State Circuits 44(12):3434–3447. https://doi.org/10.1109/JSSC.2009.2032584
21. Razavi B (2008) A millimeter-wave CMOS heterodyne receiver with On-Chip LO and divider. IEEE J Solid-State Circuits 43(2):477–485. https://doi.org/10.1109/JSSC.2007.914300
22. Floyd Ba, Reynolds SK, Pfeiffer UR, Zwick T, Beukema T, Gaucher B (2005) SiGe bipolar transceiver circuits operating at 60 GHz. IEEE J Solid-State Circuits 40(1):156–167. https://doi.org/10.1109/JSSC.2004.837250

Chapter 8
State-of-the-Art Millimeter-Wave Silicon Transceivers and Systems-on-Chip

Silicon transceiver ICs stretch across an immense array of applications. This chapter will highlight some critical aspects of transceiver design in terms of their respective application areas. The chapter concludes with a look at emerging 6G requirements and opportunities in the upper-mm-wave and THz bands.

8.1 Radar and Remote Sensing SoCs

Radar provides a cheap solution to remote sensing for a wide range of applications. Measurement of up to 100 m can be done with moderate transmit power that can easily be generated with integrated circuits. Moreover, the improvement in silicon technology has facilitated integrated radar transceivers in the mm-wave range. A large portion of the improvement in technological capability over the last decade can perhaps be attributed to the automotive industry, with advanced radar-driven safety systems a staple feature of new vehicles [1–3].

Increasing the carrier frequency can improve the azimuth resolution, as will larger antennas, array antennas, or synthetic aperture radar (SAR) techniques. However, the spatial resolution (ΔR) is ultimately limited by the transmit signal bandwidth B,

$$\Delta R = \frac{c_0}{2 \cdot B}$$

where $c_0 = 3 \times 10^8$ m/s is the speed of light [4]. The radar's resolution is defined as the minimum separation between two targets that can uniquely be distinguished. This is approximately equivalent to the 6 dB bandwidth of the transmitted signal. Moreover, the situation is complicated when different sized targets are closely spaced since the amplitude of their returns can vary significantly. In this case, the separation must be higher than ΔR. Windowing the return pulses will further degrade the

J. du Preez and S. Sinha, *State-of-the-Art of Millimeter-Wave Silicon Technology*, Lecture Notes in Electrical Engineering 945,
https://doi.org/10.1007/978-3-031-14655-8_8

resolution, even though it is necessary in most scenarios seeing that it improves detection performance [5].

8.1.1 Automotive Radar

CMOS and SiGe technologies offer unique benefits to remote sensing SoCs. For one, these systems are highly reliant on processing-intensive digital circuitry (FPGA and DSP), so good integration is a strong selling point. Moreover, multi-channel transceivers that leverage MIMO and phased array techniques have gained popularity since they enhance direction-of-arrival (DoA) estimation and detection range [6]. Arai et al. report a 77 GHz automotive radar transceiver, and its performance is summarised in Table 8.1 alongside other state-of-the-art CMOS radar SoCs.

Table 8.1 State-of-the-art radar SoCs

Reference	[6]	[7]	[8]	[9]
Technology	40 nm CMOS	45 nm CMOS	28 nm CMOS	28 nm CMOS
Building blocks	8-ch RX, 3-ch TX, PLL, digital	4-ch RX, 3-ch TX, PLL, digital	4-ch RX/TX, PLL, digital	12-ch RX, 8-ch TX, external PLL, digital
Frequency (GHz)	76 – 77	76 – 81	79	77/79
RX noise figure (dB)	8.7 – 14	18	12 (minimum)	16@80 GHz
RX P_{1dB} (dBm)	−7.4 to −10	−7	N/A	−6.7
TX power (dBm)	11.0 – 14.1	10.8	8.5	19.6 maximum (18-ch combined)
FM bandwidth (GHz)	0.3 – 1	4	N/A	N/A
Phase noise (dBc/Hz) at 1MHz offset	−91@1 MHz −116@12.5 MHz	−91	−85	−110 (ext. PLL)
Chip area (mm^2)	46.2	22	7.9	71
Power dissipation (W)	3.2	3.5	1	15
Antenna configuration	2 × 8 MIMO	N/A	2 × 2 MIMO	12 × 8 MIMO
Max. distance (m)	250	N/A	30	290

8.1.2 *Imaging*

Mm-wave and THz bands are excellent candidates for ultra-high-resolution imaging. Applications such as biomedical imaging, security systems, material analysis and quality control come to mind. Many dielectrics are entirely transparent for THz radiation, and this is non-ionising (unlike X-rays), making them safe for biomedical applications [10].

For a long time, most mm-wave and THz imagers primarily used off-chip components, resulting in bulky and expensive systems. The primary limitation to implementing these imagers on-chip is the difficulty of mm-wave and THz power generation [11]. However, the integration capability of scaled technologies provides significant benefits in cost and space, motivating a good amount of effort towards integrated solutions.

Active THz sensing is commonly based on transmission or reflection. Transmission-mode imagers are simple to design because they only require single-tone transmissions. As such, the complexity of wideband VCOs, matching networks and antennas is eliminated. Most transmission-mode imagers are multi-chip solutions, making the entire imaging alignment setup reasonably complex. On the other hand, reflection-mode imagers are single-chip, and they make use of modulated CW radars or pulsed systems. The use of modulated waveforms means that greater information depth is obtained, facilitating the possibility of high-resolution 3-D imaging [12, 13]. Such solutions do require wideband antennas and VCOs, so there is a trade-off. Additionally, pulsed radars require very fast switching schemes with minimal loss and excellent isolation, all of which is difficult to achieve on-chip in upper mm-wave and THz frequencies. The digital domain is also challenging for pulsed radars since very large processing bandwidths are needed to perform matched filtering and detection processing at baseband or digital IF [14].

Table 8.2 compares several radar imagers implemented on-chip.

8.2 Wireless Communications

8.2.1 *IEEE 802.11ad/ay WiGig*

8.2.1.1 Requirements and Challenges

The IEEE 802.11ad and 802.11ay standards target small-scale broadband wireless networks requiring multi-Gbps data rates for many users [17–19]. Four 2.16 GHz channels are allocated in the 60 GHz band for these systems, and they aim to be a significant step-up from current-generation WiFi6 and WiFi6E (IEEE 802.11ax) systems. Moreover, single-carrier systems that use QAM-16 can deliver up to 4.62 Gbps, while OFDM configurations using QAM-64 can achieve data rates as

Table 8.2 State-of-the-art imaging SoCs

Reference	[10]	[13]	[15]	[16]
Technology (f_T/f_{max})	130 nm SiGe (230/280 GHz)	130 nm SiGe (300/450 GHz)	130 nm SiGe (300/450 GHz)	130 nm SiGe (230/280 GHz)
Technique	FMCW	FMCW	FMCW	Transmission
Frequency (GHz)	168.3	227	240	321
Sensitivity (4kHz bandwidth)	87.3 fW	N/A	N/A	70.1 pW
EIRP (dBm)	16.2	30	31.6	21.1
Bandwidth (GHz/%)	27.5/16.3%	40/17.6%	60/25%	3.9/1.2%
ΔR(mm)	7.0	3.8	2.57	N/A
Chip area (mm^2)	0.72	2.25	3.2	5.14
DC power consumption (mW)	67	3500	1800	722

high as 6.76 Gbps per OFDM stream. Channel bonding and higher-order modulation schemes are needed to overcome the data rate limitations imposed by the 2.16 GHz channels, since they cannot directly support the 100 Gbps rates targeted by 802.11ay [20]. Beamforming and MIMO are other pillars of next-generation WiFi, as these together with bonding are the main improvements that 802.11ay offers over 802.11ad.

Bonding four QAM-64 channels is much easier said than done. Multiple challenging aspects such as wideband gain, low LO phase noise, precise and wideband I/Q corrections, low LO leakage and wide dynamic range each play a role in the design. 60 GHz CMOS transceivers frequently use direct-conversion architectures since these provide low power consumption and wide bandwidth [21–23]. Injection-locking techniques allow for lo phase noise and quadrature LO generation [24–26]. However, implementing wideband I/Q offset correction is a must to achieve channel bonding for QAM-64, and good solutions remain elusive. Additionally, an 8-bit 14.08 GS/s ADC is needed to sample the bonded QAM-64 signal, which implies a large time-interleaved architecture with high power consumption. In this example, with a 7.04GS/s Nyquist rate, oversampling by 2 to simplify baseband filtering results in the 14.08GS/s rate. The issue here is that an 8-bit ADC may achieve an SNDR of less than 36dB, which leaves only a small window of about 10dB for impairments.

8.2.1.2 Transceivers

Wu et al. report a 60GHz transceiver in a 65nm bulk CMOS process [20]. This transceiver uses a direct-conversion architecture for both the transmit and receive

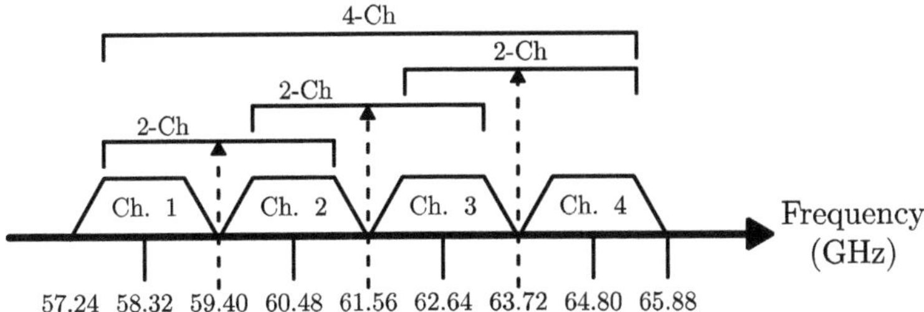

Fig. 8.1 802.11ad channel and carrier frequency allocation

paths. A mixer-first architecture is used in the transmit path to reduce power consumption further and achieve gain over a wide bandwidth. The transmitter also uses a six-stage PA with differential pre-amplifiers and a quadrature injection-locked oscillator (QILO). The receiver uses an open-loop baseband amplifier to improve linearity and bandwidth. It also contains a four-stage LNA, a QILO, I/Q current-bleeding mixers and differential RF amplifiers. The LO is comprised of a 20GHz PLL and a 60GHz QILO, so that it can generate all the required carrier frequencies with a 36/40MHz reference source. Figure 8.1 shows the allocated 802.11ad channels along with their respective carrier frequencies.

Bonding configurations include 1–2, 2–3, 3–4 and 1–2-3–4, as illustrated by Fig. 8.1.

Tomkins et al. report an 802.11ad transceiver in a 130nm SiGe BiCMOS process [27]. The transmitter implemented here uses direct conversion with differential I/Q inputs at zero IF and a single-ended output at 60GHz. A chain of switched attenuators at baseband and variable gain RF amplifiers realise gain control throughout the transmit chain. Phase and amplitude correction is placed in the LO path to implement image rejection. The receive path also implements direct conversion, and supports differential I/Q zero-IF outputs as well. The receiver also contains a VGA at baseband along with automatic gain control (AGC).

Table 8.3 summarises key results from the literature.

Based on the literature, 802.11ay applications seem to favour CMOS technologies heavily, whereas some authors report 802.11ad transceivers in SiGe BiCMOS. Silicon has a massive advantage over III-IV due to its integration capability. However, the power consumption and area efficiency of bulk and heavily scaled CMOS nodes are favoured over BiCMOS alternatives.

8.2.2 5G Mobile Communications

5G was born out of an exponential increase in user demand relative to 4G. Advanced access technologies such as beam division multiple access (BDMA) and filter bank multi-carrier (FBMC) multiple access have made the transition smoother [30]. In

Table 8.3 Summary of state-of-the-art IEEE 802.11ad/ay transceivers

Reference	[28]	[29]	[27]	[20]
Process	65nm CMOS	28nm CMOS	130nm SiGe	65nm CMOS
Data rate & modulation	21.12Gbps QAM-64/2-ch bonding 28.16Gbps QAM-16/4-ch bonding	27.8Gbps QAM-16	5.9/4.6 QAM-16	28.16Gbps QAM-16/4-ch bonding 42.24Gbps QAM-64/4-ch bonding
TX power	−4.2dBm at −26.0dB EVM	N/A	13.4 − 16.2dBm at −20.5dB EVM	8.5dBm at −21dB EVM/ 7dBm at −22dB EVM
RX noise figure (dB)	5.4	7.0	3 − 5	5.7/5.7
Core chip area (mm^2)	0.96	1.1	4.97	7.92
Building blocks	TX, RX, LO	Digital polar TX, direct-conversion RX, TX, RX, LO, PLL	Digital I/O, monitoring, TX, RX, crystal + PLL	Direct-conversion, RX, TX, LO/ Direct-conversion, 2RX, 2TX, 2LO, frequency-interleaved
Power consumption (mW)	94(TX) 105 (RX)	210(TX) 110 (RX)	53 per m link distance	251(TX) 220 (RX) / 544 (TX) 432 (RX)

a typical cellular scenario where base stations communicate with mobile stations, an orthogonal antenna beam is assigned to each mobile station and BDMA will subsequently divide that beam by location. System aspects such as latency, data rate, capacity, massive connectivity, reduced cost and consistent quality of service are not adequately addressed by 4G systems, hence the drive towards 5G and its successor, 6G. 6G is discussed in Sect. 8.3.

The requirements laid out by International Mobile Telecommunications (IMT) 2020 describe a number of usage scenarios for 5G networks [31]:

- **Improved mobile broadband (IMB)**: This aims to improve the user experience further by expanding the use cases. Scenarios such as local hotspots and wide-area coverage come to mind. High mobility and seamless coverage are desirable attributes of the wide-area coverage scenario, which is a notable improvement over 4G. In terms of hotspots, the most important characteristics are the higher total capacity to supply a larger number of users with broadband connectivity. The mobility requirement is relaxed, since only pedestrian speeds must be accommodated. On the other hand, the data rate experienced per user is much higher compared to the wide-area scenario.
- **Low-latency ultra-reliable communications (LLURC)**: Applications such as transportation safety, remote medical surgery, public protection, disaster relief and tactile internet place strong emphasis on reliability, availability and latency.
- **Massive machine communications (MMC)**: In these scenarios, a large number of cheap devices communicate in a non-latency-essential manner. As such, devices must be optimised for battery life and cost. Potential applications are smart factories that extend Industry 4.0.

Table 8.4 lists the performance requirements discussed above as outlined in IMT 2020 [31–33].

8.2.2.1 RF Technologies for 5G

The system-level requirements for 5G performance bleed into component and subsystem specifications. As such, the RF requirements can be described as follows [31, 34–36]:

- **Wideband operation**: 4G and prior cellular systems operate below 6GHz, with the bulk of real-world deployments living below 3GHz. It is no secret that spectrum is massively congested in this range, which is one of the primary motivators for extending operation into mm-wave bands (28 − 300GHz). In these bands, much wider carrier bandwidths are available, translating into better user experienced bandwidth. 5G deployments can also be split into two layers: a macro layer which is served by microwave links that handle user and control plane traffic, and a dense microlayer, served by mm-wave links that carry only user plane traffic.
- **Massive MIMO base stations**: Increasing the carrier frequency opens up opportunities for significantly larger array antennas. These antennas serve to provide

Table 8.4 5G performance requirements of IMT 2020

KPI	Use cases	Requirements
Peak data rate	IMB	DL: 20Gbps UL:10Gbps
Experienced data rate (Dense urban scenario)	IMB	DL: 100Mbps UL: 50Mbps
Peak spectral efficiency	IMB	DL: 30bps/Hz UL:15bps/Hz
5% Experienced spectral efficiency	IMB	Indoor hotspot: DL: 0.3bps/Hz, UL: 0.21bps/Hz Dense urban: DL: 0.225bps/Hz, UL: 0.15bps/Hz Rural: DL: 0.12bps/Hz, UL: 0.045bps/Hz
Average spectral efficiency	IMB	Indoor hotspot: DL: 9bps/Hz/TRxP[a] UL: 6.75bps/Hz/TRxP Dense urban: DL: 7.8 bps/Hz/TRxP, UL: 5.4bps/Hz/TRxP Rural: DL: 3.3bps/Hz/TRxP, UL:1.6bps/Hz/TRxP
User latency	IMB, LLURC	LLURC: 1ms IMB:4ms
Control latency	IMB, LLURC	20ms for both IMB and LLURC
Connection density	MMC, IMB	10^6devices/km^2
Energy efficiency	MBB, MMC	Support for extended sleep mode to facilitate low power consumption during absence of data transmission, no hard requirement
Reliability	LLURC	$1 - 10^{-5}$ probability of successfully transmitting 32 byte layer 2 data unit within 1ms at coverage edge
Mobility	IMB	Maximum of 500km/h
Bandwidth	IMB	100MHz – 1 GHz(1GHz for carriers above 6GHz)

[a]TRxP: transmission layer reception points, which can vary from small-, macro-, femto- and picocells as well as relay nodes and radio heads [34]

high gain through spatial multiplexing in order to combat higher path loss. For mm-wave 5G bands, arrays with anything between 256 and 1024 antennas are feasible. Moreover, each array can consist of sub-arrays and utilise cross-polarised elements to provide polarisation diversity. Massive MIMO also enables transmission to multiple users scattered in azimuth and elevation, and the number of spatial streams limits the number of simultaneous users that can be serviced at the base station end.

- **Denser networks**: Increasing network density enables offloading traffic to smaller cells, alleviating the traffic that a single cell must be able to handle. This is especially useful in dense urban environments and indoor hotspots. However, macrocell coverage via microwave links is still required to handle control plane data, and increasing density also increases interference at the cell edge. As such, cells cannot be shrunk ad infinitum. Cooperative scheduling, for example, can be used to mitigate intercell interference to an extent.
- **New waveforms**: Some 5G applications require re-thinking waveforms, multiple-access schemes and modulation. For example, large-scale M2M communications use very small packet sizes, in which case the overhead of OFDM (scheduling, resource allocation, etc.) will result in inefficient communication. Scheduled transmission also induces overhead and latency due to the request-response nature of operation.

8.2.2.2 Transceivers

Kim et al. report a 28GHz direct-conversion transceiver packaged with a 2×4 antenna array [37]. The direct-conversion topology is preferred because it provides good linearity and noise figure, and it simplifies integrating the complete RX and TX chains on-chip. On the other hand, sliding-IF and superheterodyne topologies can be advantageous since they use low LO frequencies, can implement power combining and splitting at IF, and perform better in terms of I/Q mismatch. However, the conversion chain is significantly longer, which contributes to system loss and degrades dynamic range and EVM. Table 8.5 summarises the performance of several designs from literature across a variety of processes and architectures.

8.3 6G and Future Mm-Wave Systems

8.3.1 Evolution from 5 to 6G

Wireless communication systems have evolved tremendously in recent years. Many stakeholders (standards bodies, end-users, research institutions and commercial service providers) have benefitted tremendously from the technologies that accompanied 5G. Some of these technologies include massive MIMO, virtual reality (VR), machine-to-machine (M2M) communication, contactless payments and mobile broadband, and all have shown the potential of 5G [41]. Today even, 5G technologies continue to evolve and improve products and services.

As the needs of society continue to evolve, a number of applications and use cases cannot be adequately implemented within the current set of 5G technologies. Holographic teleportation—which will likely form part of the next generation of

VR—requires Tbps data rates and sub-microsecond latencies, surpassing the capability of even the mm-wave 5G bands [42, 43]. Another issue is the substantial increase in automation that has become a staple of many industries. The migration from Industry 4.0 to the newer Industry X.0 paradigm will likely utilise connectivity density far beyond the capability of current 5G networks [44]. As such, a rework of the current generation of network management practices will likely be necessary, as well as significant improvements in energy efficiency.

Table 8.6 highlights essential differences in key performance indicators (KPIs) between 5 and 6G systems [45].

Table 8.6 highlights the fundamental metrics by which system performance can be evaluated, starting with System Capacity.

Table 8.5 Beamforming transceivers for mm-wave 5G

Reference	[37]	[38]	[39]	[40]
Technology	28nm CMOS	180nm SiGe BiCMOS	0.15μm GaN HEMT (PA)/GaAs (LNA, mixers, switches)	130nm SiGe BiCMOS
Architecture	Direct-conversion, RF beamforming	Direct-conversion, RF beamforming	Direct IF (multi-chip), digital beamforming	Double-conversion sliding-IF, RF beamforming
Frequency (GHz)	$25.8 - 28.0$	$27.8 - 31.2$	28	$27 - 29$
Beamforming elements	2×4 TRX	4 per TRX front-end, $2 \times 4 \times 4 = 32$ per transceiver	15 per transceiver, 480 (32 transceivers)	16 per transceiver, 64 (4 transceivers)
RX noise figure (dB)	6.7 at $G_P = 69$dB 13.6 at $G_P = 30$dB	7.5	N/A	≥ 6.0(PA off)
TX EIRP (dBm)	31.5 at $P_{sat} = 9.5$dBm 24.0 at $P_{out} = 2.0$dBm (i.e. 7.5dB backoff)	$P_{1\text{dB}} = 41$dBm with 32 elements (Single PA: $P_{1\text{dB}} = 11.7$dBm)	68 with 480 elements, $P_{out} = 42$dBm (7dB backoff)	28 (16 elements)
EVM at medium power	RX: 2.15% at $P_{in} = -60$dBm TX: 2.20% at $P_{in} = 0$dBm (QAM-64, LTE 20MHz)	7.5% (QAM-64)	4.5% (QAM-64)	Not reported, QAM-64 tested
DC power consumption (mW)	RX: 400 TX: 680	RX: 130 TX: 200	450×10^3	RX: 3300 TX: 4600
Die size	2.8×2.6mm^2	2.5×4.7mm^2 per 4-element TRX front-end	Not reported	15.6×10.6mm^2

Table 8.6 Comparison between 5 and 6G system requirements

KPI	5G	6G
Capacity		
Peak data rate (Gbps)	10 – 20	> 1000
Experienced data rate (Gbps)	0.1	1
Peak spectral efficiency (b/s/Hz)	30 3 – 5× 4G	60 > 3× 5G
Experienced spectral efficiency (b/s/Hz)	0.3	3
Peak channel bandwidth (GHz)	1	100
Area traffic capacity (Mbps/m^2)	10	1000
Connection density (devices/km^2)	10^6	10^7
Latency		
End-to-end latency (ms)	1	
Jitter (ms)	N/A	10^{-3}
Management		
Energy efficiency (Tb/J)	N/A	1
Reliability (packet error rate)	10^{-5}	10^{-9}
Mobility (km/h)	300 – 500	> 1000

- **Capacity**: capacity KPIs are primarily focused on the throughput of the system as a whole. The experience data rate and spectral efficiency specify the minimum guaranteed performance in 95% of all service locations.
- **Latency**: End-to-end latency and jitter requirements are stringent for 6G systems, especially since there are no jitter specifications for 5G.
- **Management**: Specifications relating to the management and orchestration of networks are captured in Table 8.6. 6G is the first class of wireless system with an energy efficiency requirement attached to it.

Identifying the requirements of future networks is the first step towards their realisation. Next, determining the evolutionary technologies needed across multiple domains of wireless communications is appropriate. With 6G, there are multiple breakthrough innovations required:

- **Spectrum usage and radio design improvements**: 5G ensured the mainstream adoption of mm-wave wireless networks. The increase in bandwidth and spectral efficiency will inevitably require radios operating in THz and sub-THz bands above the mm-wave region. Spectrum challenges that are omnipresent in the process of pushing the boundaries of operating frequencies will have to be solved as well.
- **Network architectures**: A major paradigm shift in network architecture that 6G may require is moving away from the classic cell-based model. This classic architecture cannot cope with the connection density and area traffic capacity

required by 6G. This will also create significant challenges in the deployment of network infrastructure.

- **Increase in intelligence and automation**: Manual network configuration will likely become a thing of the past, given the immense latency and reliability requirements associated with 6G. Network intelligence will become central to 6G systems.
- **Extra-terrestrial coverage**: Achieving wireless ubiquity will require 6G to extend outside of terrestrial networks. As such, deep-space and near-earth connectivity will become part of the network paradigm.

8.3.2 THz-Band Communication

THz band communication (100GHz–10THz) is a key research area in wireless communications. THz bands can provide multi-Tbps links for numerous applications. The FCC recently released several bands above 95GHz for research purposes [41]. Network operators have slowly adopted lower mm-wave bands (e.g. the 28 and 39GHz bands) for 5G cellular communications, anticipating that data rates will reach 100Gbps. Real-world testing has been much less impressive, with actual speeds around 1Gbps [45]. Many factors contribute to this, such as the complexity of accurately modelling the channel of a 5G link, circuit design, manufacturing imperfections and interference. While THz bands have been used for ultra-high-resolution imaging [46, 47] and spectroscopy [11], among others, wireless communication has yet to gain significant interest from the research community. The abundance of spectrum between upper mm-wave and infrared bands is deemed a "no-man's land", especially due to the historical lack of transceivers and antennas operating there.

The unique photonics and electromagnetic characteristics at such high frequencies open up unique opportunities. In addition to Tbps cellular networks, THz spectrum can be utilised for further applications [41, 44, 48–50]:

- **Local area networks**: Short-range links (below 10m) operating in the 625 – 725GHz and 780 – 910GHz bands are feasible due to THz-optics bridges that facilitate a seamless transition from wireless THz links to fibre-optic backhaul.
- **Personal area networks**: Data rates akin to fibre broadband are possible without requiring physical connections up to a few meters. Applications could be found in small offices, boardrooms and multimedia kiosks.
- **Wireless network-on-chip (WNoC)**: Transceiver specifications are increasingly driving more integrated designs and miniaturisation. THz links can potentially facilitate the interconnection between multiple wireless modules within an enclosure as a means of replacing the wired connections in current-generation systems.
- **Nano-cell networks**: THz frequencies are better suited to nano networks than mm-waves because of their short wavelengths. Interconnected nano-devices and machines with excellent spatial frequency re-use are a possibility.

- **Satellite communication**: Communication links between satellites are not constrained by atmospheric attenuation. As such, THz bands are excellent options for inter-satellite links. THz bands also do not require the strict beam alignment of optical links, which could mean that the links are much more stable when satellites drift outside their orbit.

8.3.3 *THZ Devices*

Increased output power, better sensitivity, lower phase noise are all key requirements for THz transceivers that drive the efforts towards improving performance. THz-band signal generation can be separated into electronic-based, photonics-based and emerging material-based techniques [51]. Photonics favours III-IV technologies such as GaAs and InP because of their high electron mobility. These devices generated pulse widths in the femtosecond range, and indoor experiments have shown 50Gbps data rates at 300GHz with uni-travelling-carrier photodiodes [52]. Newer devices will likely depend on the development of novel materials such as graphene and carbon nanotubes. At room temperature, these materials have electron mobilities ranging from 8000 to $10000\text{cm}^2/\text{V} \cdot \text{s}$, significantly more than the $1400\text{cm}^2/\text{V} \cdot \text{s}$ of silicon and comparable to the $8500\text{cm}^2/\text{V} \cdot \text{s}$ of GaAs. Graphene-based devices offer exceptional electrical, optical and mechanical properties, and power detectors at 600GHz have been demonstrated [53]. These devices offer excellent potential for even beyond THz operation.

References

1. Bloecher HL, Dickmann J, Andres M (2009) Automotive active safety & comfort functions using radar. In: Proceedings—2009 IEEE International Conference on Ultra-Wideband, ICUWB 2009, vol 2009, pp 490–494. https://doi.org/10.1109/ICUWB.2009.5288790
2. Wenger J (2005) Automotive radar—status and perspectives. In: IEEE Compound Semiconductor Integrated Circuit Symposium, pp 21–24. https://doi.org/10.1109/CSICS.2005.1531741
3. Strohm KM, Bloecher H-L, Schneider R, Wenger J (2005) Development of future short range radar technology. European Radar Conference, 2005. EURAD 2005, pp 185–188. https://doi.org/10.1109/EURAD.2005.1605591
4. Richards MA, Scheer JA, Holm WA (2010) Principles of modern radar—basic principles. Scitech Publishing, Edison, New Jersey
5. Harris FJ (1978) On the use of windows for harmonic analysis with the discrete Fourier transform. Proc IEEE 66(1):51–83. https://doi.org/10.1109/PROC.1978.10837
6. Arai T et al (2021) A 77-GHz 8RX3TX transceiver for 250-m long-range automotive radar in 40-nm CMOS technology. IEEE J Solid-State Circuits 56(5):1332–1344. https://doi.org/10.1109/JSSC.2021.3050306
7. Ginsburg BP et al (2018) A multimode 76-to-81GHz automotive radar. In: IEEE Solid-State Circuits Conference (ISSCC), pp 158–160
8. Guermandi D et al (2017) A 79-GHz 2 × 2 MIMO PMCW Radar SoC in 28-nm CMOS. IEEE J Solid-State Circuits 52(10):2613–2626. https://doi.org/10.1109/JSSC.2017.2723499

9. Giannini V et al (2019) A 192-virtual-receiver 77/79GHz GMSK code-domain MIMO radar system-on-Chip. In: IEEE International Solid-State Circuits Conference (ISSCC), pp 164–166
10. Mostajeran A, Cathelin A, Afshari E (2017) A 170-GHz fully integrated single-chip FMCW imaging radar with 3-D imaging capability. IEEE J Solid-State Circuits 52(10):2721–2734. https://doi.org/10.1109/JSSC.2017.2725963
11. Khatibi H, Afshari E (2017) Towards efficient high power mm-wave and terahertz sources in silicon: one decade of progress. SiRF 2017—2017 IEEE 17th Topical Meeting on Silicon Monolithic Integrated Circuits in RF Systems, pp 4–8. https://doi.org/10.1109/SIRF.2017.7874355
12. Marimuthu J, Bialkowski KS, Abbosh AM (2016) Software-defined radar for medical imaging. IEEE Trans Microw Theory Tech 64(2):643–652. https://doi.org/10.1109/TMTT.2015.2511013
13. Thomas S, Bredendiek C, Jaeschke T, Vogelsang F, Pohl N (2016) A compact, energy-efficient 240 GHz FMCW radar sensor with high modulation bandwidth. In: GeMiC 2016—2016 German Microwave Conference, pp 397–400. https://doi.org/10.1109/GEMIC.2016.7461639
14. Richards MA (2005) Fundamentals of radar signal processing. McGraw-Hill, New York City, New York
15. Grzyb J, Statnikov K, Sarmah N, Heinemann B, Pfeiffer UR (2016) A 210–270-GHz circularly polarised FMCW radar with a single-lens-coupled SiGe HBT Chip. IEEE Trans Terahertz Sci Technol 6(6):771–783. https://doi.org/10.1109/TTHZ.2016.2602539
16. Jiang C et al (2016) A fully integrated 320 GHz coherent imaging transceiver in 130 nm SiGe BiCMOS. IEEE J Solid-State Circuits 1–14
17. Zhou P et al (2018) IEEE 802.11ay-based mmWave WLANs: design challenges and solutions. IEEE Commun Surv Tutor 20(3):1654–1681. https://doi.org/10.1109/COMST.2018.2816920
18. Ghasempour Y, Da Silva CRCM, Cordeiro C, Knightly EW (2017) IEEE 802.11ay: next-generation 60 GHz communication for 100 Gb/s Wi-Fi. IEEE Commun Mag 55(12):186–192. https://doi.org/10.1109/MCOM.2017.1700393
19. Perahia E, Cordeiro C, Park M, Yang LL (2010) IEEE 802.11ad: defining the next generation. In: 7th IEEE consumer communications and networking conference, pp 1–5. https://doi.org/10.1109/CCNC.2010.5421713
20. Wu R et al (2017) 64-QAM 60-GHz CMOS Transceivers for IEEE 802.11ad/ay. IEEE J Solid-State Circuits 52(11):2871–2891. https://doi.org/10.1109/JSSC.2017.2740264
21. Tomkins A, Aroca RA, Yamamoto T, Nicolson ST, Doi Y, Voinigescu SP (2009) A zero-IF 60 GHz 65 nm CMOS transceiver with direct BPSK modulation demonstrating up to 6 Gb/s data rates over a 2 m wireless link. IEEE J Solid-State Circuits 44(8):2085–2099. https://doi.org/10.1109/JSSC.2009.2022918
22. Okada K et al (2011) A 60GHz 16QAM/8PSK/QPSK/BPSK direct-conversion transceiver for IEEE802.15.3c. IEEE J Solid-State Circuits 46(12):2988–3004. https://doi.org/10.1109/ASICON.2011.6157381
23. Saito N et al (2013) A fully integrated 60-GHz CMOS transceiver chipset based on WiGig/IEEE 802.11ad with built-in self calibration for mobile usage. IEEE J Solid-State Circuits 48(12):3146–3159. https://doi.org/10.1109/JSSC.2013.2279573
24. Kishimoto S, Maruhashi K, Ito M, Morimoto T, Hamada Y (2005) A 60-GHz-band subharmonically injection locked VCO MMIC operating over wide temperature range. In: IEEE MTT-S international microwave symposium digest, pp 5–8
25. Liao D, Zhang Y, Dai FF, Chen Z, Wang Y (2020) An mm-wave synthesizer with robust locking reference-sampling PLL and wide-range injection-locked VCO. IEEE J Solid-State Circuits 55(3):536–546. https://doi.org/10.1109/JSSC.2019.2959513
26. Li A, Zheng S, Yin J, Luo X, Luong HC (2014) A 21–48 GHz subharmonic injection-locked fractional-N frequency ssynthesiser for multiband point-to-point backhaul communications. IEEE J Solid-State Circuits 49(8):1785–1799. https://doi.org/10.1109/JSSC.2014.2320952
27. Tomkins A et al (2015) A 60 GHz, 802.11ad/WiGig-compliant transceiver for infrastructure and mobile applications in 130 nm SiGe BiCMOS. IEEE J Solid-State Circuits 50(10):2239–2255. https://doi.org/10.1109/JSSC.2015.2436900

28. Pang J et al (2020) A 28.16-Gb/s area-efficient 60-GHz CMOS bidirectional transceiver for IEEE 802.11ay. IEEE Trans Microw Theory Tech 68(1):251–262. https://doi.org/10.1109/TMTT.2019.2938160
29. Dasgupta K et al (2018) A 60-GHz transceiver and baseband with spolarisation MIMO in 28-nm CMOS. IEEE J Solid-State Circuits 53(12):3613–3627. https://doi.org/10.1109/JSSC.2018.2876473
30. Gupta A, Jha RK (2015) A survey of 5G network: architecture and emerging technologies. IEEE Access 3:1206–1232. https://doi.org/10.1109/ACCESS.2015.2461602
31. Shafi M et al (2017) 5G: a tutorial overview of standards, trials, challenges, deployment, and practice. IEEE J Sel Areas Commun 35(6):1201–1221. https://doi.org/10.1109/JSAC.2017.2692307
32. Mezzavilla M et al (2018) End-to-end simulation of 5G mmWave networks. IEEE Commun Surv Tutor 20(3):2237–2263. https://doi.org/10.1109/COMST.2018.2828880
33. Boccardi F, Heath R, Lozano A, Marzetta TL, Popovski P (2014) Five disruptive technology directions for 5G. IEEE Commun Mag 52(2):74–80. https://doi.org/10.1109/MCOM.2014.6736746
34. Fuentes M et al (2020) 5G new radio evaluation against IMT-2020 key performance indicators. IEEE Access 8:110880–110896. https://doi.org/10.1109/ACCESS.2020.3001641
35. Uwaechia AN, Mahyuddin NM (2020) A comprehensive survey on millimeter wave communications for fifth-generation wireless networks: feasibility and challenges. IEEE Access 8:62367–62414. https://doi.org/10.1109/ACCESS.2020.2984204
36. Gangakhedkar S, Cao H, Ali AR, Ganesan K, Gharba M, Eichinger J (2018) Use cases, requirements and challenges of 5G communication for industrial automation. In: IEEE international conference on communications workshops, ICC workshops 2018—proceedings, pp 1–6. https://doi.org/10.1109/ICCW.2018.8403588
37. Kim H-T et al (2018) A 28-GHz CMOS direct conversion transceiver with packaged 2x4 antenna array for 5G cellular system. IEEE J Solid-State Circuits 53(5):1245–1259. https://doi.org/10.1109/jssc.2018.2817606
38. Kibaroglu K, Sayginer M, Rebeiz GM (2017) An ultra low-cost 32-element 28 GHz phased-array transceiver with 41 dBm EIRP and 1.0–1.6 Gbps 16-QAM link at 300 meters. In: IEEE Radio Frequency Integrated Circuits Symposium (RFIC), pp 73–76. https://doi.org/10.1109/RFIC.2017.7969020
39. Kuwabara T, Tawa N, Tone Y, Kaneko T (2017) A 28 GHz 480 elements digital AAS using GaN HEMT amplifiers with 68 dBm EIRP for 5G long-range base station applications. In: IEEE compound semiconductor integrated circuit symposium (CSICS), pp 1–4. https://doi.org/10.1109/CSICS.2017.8240471
40. Sadhu B et al (2017) A 28GHz 32-element phased-array transceiver IC with concurrent dual polarized beams and 1.4° beam-steering resolution for 5G communication. In: IEEE international solid-state circuits conference, pp 128–130. https://doi.org/10.1109/ISSCC.2017.7870294
41. Tataria H, Shafi M, Molisch AF, Dohler M, Sjoland H, Tufvesson F (2021) 6G wireless systems: vision, requirements, challenges, insights, and opportunities. Proc IEEE 109(7):1166–1199. https://doi.org/10.1109/JPROC.2021.3061701
42. Wells J (2009) Faster than fiber: the future of multi-G/s wireless. IEEE Microwave Mag 10(3):104–112. https://doi.org/10.1109/MMM.2009.932081
43. Jornet JM, Akyildiz IF (2010) Channel capacity of electromagnetic nanonetworks in the terahertz band. IEEE Int Conf Commun. https://doi.org/10.1109/ICC.2010.5501885
44. Rappaport TS et al (2019) Wireless communications and applications above 100 GHz: opportunities and challenges for 6g and beyond. IEEE Access 7:78729–78757. https://doi.org/10.1109/ACCESS.2019.2921522
45. Akyildiz IF, Kak A, Nie S (2020) 6G and beyond: the future of wireless communications systems. IEEE Access 8:133995–134030. https://doi.org/10.1109/ACCESS.2020.3010896
46. Statnikov K, Grzyb J, Heinemann B, Pfeiffer UR (2015) 160-GHz to 1-THz multi-color active imaging with a lens-coupled SiGe HBT chip-set. IEEE Trans Microw Theory Tech 63(2):520–532. https://doi.org/10.1109/TMTT.2014.2385777

47. Appleby R, Anderton RN (2007) Millimeter-wave and submillimeter-wave imaging for security and surveillance. Proc IEEE 95(8):1683–1690. https://doi.org/10.1109/JPROC.2007.898832
48. Akyildiz IF, Jornet JM, Han C (2014) TeraNets: ultra-broadband communication networks in the terahertz band. IEEE Wirel Commun 21(4):130–135. https://doi.org/10.1109/MWC.2014.6882305
49. Song HJ, Nagatsuma T (2011) Present and future of terahertz communications. IEEE Trans Terahertz Sci Technol 1(1):256–263. https://doi.org/10.1109/TTHZ.2011.2159552
50. Voinigescu SP, Shopov S, Bateman J, Farooq H, Hoffman J, Vasilakopoulos K (2017) Silicon millimeter-wave, terahertz, and high-speed fiber-optic device and benchmark circuit scaling through the 2030 ITRS horizon. Proc IEEE 105(6):1087–1104. https://doi.org/10.1109/JPROC.2017.2672721
51. Kenneth KO, Chang MCF, Shur M, Knap W (2009) Sub-millimeter wave signal generation and detection in CMOS. In: IEEE MTT-S international microwave symposium digest, pp 185–188. https://doi.org/10.1109/MWSYM.2009.5165663
52. Nagatsuma T, Carpintero G (2015) Recent progress and future prospect of photonics-enabled terahertz communications research. IEICE Trans Electron E98.C(12):1060–1070. https://doi.org/10.1587/transele.E98.C.1060
53. Zak A et al (2014) Antenna-integrated 0.6 THz FET direct detectors based on CVD graphene. Nano Lett 14(10):5834–5838. https://doi.org/10.1021/nl5027309

www.ingramcontent.com/pod-product-compliance
Ingram Content Group UK Ltd.
Pitfield, Milton Keynes, MK11 3LW, UK
UKHW021834270726
14058UKWH00001B/147

* 9 7 8 3 0 3 1 1 4 6 5 7 2 *